Praise for Jill Heinerth's
Into the Planet

"Breathtaking. . . . Written in cinematic detail, *Into the Planet* is a thrilling portrait of bravery, innovation, and the extreme limits of human capability. . . . One of the most hair-raising accounts of extreme exploration I've read in recent memory."

—*Gizmodo*

"Superb, honest, incredibly engaging. . . . A captivating biography and a love letter to a sport where any small mistake can result in death—and any perfect dive can mean an amazing discovery."

—NPR

"Exciting, edge-of-your-seat reading. . . . Bursts with full-throttle exuberance for the highs, and sometimes even the lows, of being a pioneering, modern-day explorer."

—*BookPage* (starred review)

"A hugely enjoyable, inspiring, and thrilling self-exploration."

—*Diver*

"An eye-opening, edge-of-the-seat, thrill of a read that's destined to become a trailblazing classic of diving literature."

—Nektonix

"A breathtaking thrill ride into a deeper world that exists within this one. Few people have witnessed the sublime beauty of labyrinthine underwater caves or the inside of an iceberg—to venture in for even a glimpse is to put your life at risk—but Jill Heinerth knows these realms intimately. Her courage and heart are as evident in her writing as they are in her groundbreaking explorations. You will not be able to put this book down."

"A meditation on the paradoxical power of fear."

"Enchanting. . . . An exhilarating, deeply personal memoir. . . . Told with sensitivity and joyful enthusiasm, this is an inspiring story that will appeal to many, especially adventurous young women."

"Every form of extreme endeavor produces a preeminent practitioner: Messner in mountains, Honnold on rock, Ballard in oceans. As *Into the Planet* eloquently demonstrates, for cave diving it may well be Jill Heinerth. Her newest book chronicles, in chiseled prose, a lifetime pursuing science and discovery in exploration's most lethal and least forgiving discipline. Time and again, Heinerth has ventured into realms where mishaps can, quite literally, take your breath away. Just reading about her exploits can do the same—happily with less final results. And there is this: like all cave divers, Heinerth operated in the 'overhead environment'—a ceiling of solid rock. Unlike them, she had to break through another overhead environment that was, if less deadly, certainly more insidious: a glass ceiling of gender-based prejudice and animus. The fact that she prevailed against both makes her exploits even more remarkable and her story all the more compelling."

"Gripping."

"Exhilarating. . . . Heinerth offers a fish's-eye view of the terrifying beauty of the deep blue sea."

"I can guarantee you one thing: Jill Heinerth will not look back on her life with any regrets about a lack of bravery or passion. In this gripping, life-of-gusto story, we plunge into Heinerth's eccentric world of death-defying caves, and into her inner sanctum forged by the thrill of discovery. I read wide-eyed, page after page."

"A thrill ride into unfamiliar worlds."

"Jill Heinerth's lifelong passion to explore the most hidden recesses of our planet has driven her to boldly go where—quite literally—no one had gone before. She takes us along on her awe-inspiring journey, revealing the oceans of opportunity that exist right beneath our very feet, if we are willing to push the limits of conventional knowledge and explore what most will never see."

"A compelling story of an adventurous, bold life yielding rewards of love, discovery, and personal growth."

"This must-read memoir looks back at an amazing career and provides insight into parts of the world that few of us will ever see in person."

—*EcoWatch*

"By turns terrifying, exhilarating, and inspirational, *Into the Planet* celebrates the fine madness of deep-sea cave diving. Talk about extreme adventure. If you don't know the name Jill Heinerth, prepare to be dazzled and astonished. Move over, Jon Krakauer. This well-written memoir is destined to become a world classic of exploration literature."

—Ken McGoogan, author of *Fatal Passage*

"*Into the Planet* takes us into the risk-laden, invisible world beneath our feet. Jill Heinerth's journey from the office to the far reaches of the planet reveals how the physical difficulties often pale in comparison to the personal challenges related to maintaining relationships, overcoming gender bias, and honestly assessing one's own tolerance of danger."

—Kenny Broad, PhD, National Geographic Explorer

"Cave divers are another breed, which I never truly understood—that is, until I met Jill Heinerth on a cave diving trip off Santa Cruz Island and had a chance to read her new book *Into the Planet*, which helped to explain it all. A must-read!"

—Robert D. Ballard, PhD, discoverer of the RMS *Titanic*

INTO

THE

PLANET

INTO

THE

PLANET

MY LIFE AS A CAVE DIVER

JILL HEINERTH

ecco
An Imprint of HarperCollinsPublishers

HarperCollins books may be purchased for educational, business, or sales promotional use. For information, please email the Special Markets Department at SPsales@harpercollins.com.

Published in Canada in 2019 by Doubleday Canada.

A hardcover edition of this book was published in 2019 by Ecco, an imprint of HarperCollins Publishers.

FIRST ECCO PAPERBACK EDITION PUBLISHED 2020

Library of Congress Cataloging-in-Publication Data has been applied for.

ISBN 978-0-06-269155-2 (pbk.)

20 21 22 23 24   LSC   10 9 8 7 6 5 4 3 2

To my husband, Robert McClellan, who supports and celebrates my career even though it scares him.

*"Only those who will risk going too far
can possibly find out how far one can go."*

T.S. ELIOT, from the preface to *Transit of Venus:
Poems* by Harry Crosby (1931)

PROLOGUE

2001

IF I DIE, it will be in the most glorious place that nobody has ever seen.

I can no longer feel the fingers in my left hand. The glacial Antarctic water has seeped through a tiny puncture in my formerly waterproof glove. If this water were one-tenth of a degree colder, the ocean would become solid. Fighting the knife-edged freeze is depleting my strength, my blood vessels throbbing in a futile attempt to deliver warmth to my extremities.

The archway of ice above our heads is furrowed like the surface of a golf ball, carved by the hand of the sea. Iridescent blue, Wedgwood, azure, cerulean, cobalt, and pastel robin's egg meld with chalk and silvery alabaster. The ice is vibrant, bright, and at the same time ghostly, shadowy. The beauty contradicts the danger. We are the first people to cave dive inside an iceberg. And we may not live to tell the story.

It's February, in the middle of what passes for summer in Antarctica. My job, for National Geographic, is to lead an advanced technical diving team in search of underwater caves deep within the largest moving object on earth, the B-15 iceberg. I had known that diving into tunnels inside this giant piece of ice would be difficult, but I hadn't calculated that getting out would be nearly impossible. The tidal currents accelerated so quickly that they've caged us inside the ice. We're trapped in this frozen fortress, and I have to figure out how to escape.

There are no training manuals or protocols to follow. When you're the first to do something, there's nobody to call for help. The most qualified cave-diving team in the world, with the experience and skills to rescue us, is right here, trapped inside the B-15 iceberg: my husband, Paul Heinerth, our close friend, Wes Skiles, and me.

The glazed tunnel we're swimming through is magnificent. Three hundred feet of ice presses down upon us from above this narrow passage, groaning with emphatic creaks and pops that signal its instability. The current is gaining momentum, and the garden of life on the seafloor beneath the iceberg bends like palm trees in a hurricane. Frilly marine creatures—brilliant orange sponges, worms that look like Christmas trees, and vibrant red stalks—double over and shake in the flow of the tide. Wes is trailing behind Paul and me, attempting to film our exploration for National Geographic, and I sense him losing ground in the current.

Our planned one-hour dive is stretching out of control, and I'm not sure how long we can tolerate the cold. Can we survive two hours? Three? The fifteen crewmates on our battered research vessel *Braveheart* are likely unaware of the drama unfolding in the water. They only know that we're overdue. If we don't return soon, our captain will have to call for help into a radio handset, but no one will hear him. We're beyond the range of communications—utterly alone against the wilderness. And there are no other capable divers on board. Our colleagues will search the horizon through binoculars; they'll launch the ship's helicopter and ferret feverishly over the endless white ice of the Ross Sea. But they'll know that nobody survives for long in these indifferent waters. We would be remembered at best as gutsy, but more likely as lunatics.

The incredible pain in my hand begins to yield to a numbness that threatens to hijack my resolve. I know that as my core temperature drops, confusion will soon follow. When pain subsides, death is often lurking. I plunge my frostbitten hand into the doughy seafloor to pull myself forward, and columns of clay rise like smoke. I'm simultaneously hot and cold. My chest is heaving, my lungs burning.

There's a beam of daylight, soft and elusive, about three hundred yards away, and I begin kicking as hard as I can, latching on to anything on the ocean floor that could edge me closer to it. I can hear Paul's and Wes's heavy panting, but my mind is turning inward to my own survival as I gain one inch of ground at a time.

How does a dying person know when it's over? They say your life flashes before your eyes, but that isn't happening to me now. All I can think about is escaping from the water that I love more than anything else. I've spent my life immersed in a relationship with this element that nourishes and destroys, buoys and drowns— that has both freed me and taken the lives of my friends. Now, I have come to my moment of reckoning. My life began in water, and I refuse to accept that it may end here.

I'M GOING TO carry you on a journey to places you've never imagined, deeper inside underwater caves than any woman has ventured. I will take you on an uncomfortable rendezvous with fear. You will feel cold and claustrophobic when you read this book. But I challenge you to recognize the humanity in that sensation of terror you're experiencing. I encourage you to accept that you are an explorer like me.

As young children, we exhibit a complete wonderment and lack of fear about the world around us. In our youth, everything is fresh and sensational, and we don't have to work hard at exploring. For me, it seemed normal to examine the natural world around me, boldly cresting every hill without fear of what was on the other side. But eventually, painful life experiences shaped me to accept fear as a reigning doctrine. Like most people, by adulthood I found myself searching for stability and certainty. It's easy to become comfortable with the status quo, more concerned about losing ground than reaching new heights.

We all have fears, rational and irrational—spiders, storms and germs, killer bees, killer dogs, and killer cars with runaway accelerators. We worry about losing our jobs, paying our bills, and protecting our families. Government leaders encourage us to fear others, especially when they don't look like or worship like us. Nations shut their borders and citizens close their doors, choosing to be sequestered in triple-locked, protected enclaves and preferring synthetic lives where stimulation is satiated digitally. All news is breaking news these days, and sometimes it seems like every headline promises to march us closer to the end of the world as we know it.

For me, acknowledging and then living my wildest dream has meant learning to accept and welcome fear. Cave diving, at the intersection of earthbound science, exploration, and discovery, tests the extremes of human capability. My job is never simple—whether I'm filming documentary footage, mapping previously undiscovered caves, or gathering data and specimens for a scientific mission, I'm also always combating the elements, navigating tight passages, and monitoring the complex life-support system that keeps me breathing underwater. My survival depends on my balancing fear and confidence. When I get snagged in a tight, body-contorting crevice or am lost in the blinding murk of a silt-out, I have to measure each setback in single breaths. If I allow fear to seize me, then my breathing shoots through the roof at a time when every molecule of oxygen I use up tugs me closer to death.

It is impossible, at such times, not to recall that more people have died exploring underwater caves than climbing Mount Everest. Perhaps more than in any other adventure endeavor. Cave diving is so risky that even the most casual enthusiasts can't get life insurance at any price. Even with modern equipment and thorough training, an average of twenty people drown every year in these watery catacombs.

So why would any reasonably sane person want to swim into what many would think of as a death trap? For me, cave diving is a kind of "back to the womb" experience. It feels primal, like I'm being called by some ancient ancestor. Underwater caves are alluring, challenging, and seductively dangerous. They're often filled with gin-clear water and fantastic geologic formations unlike anything seen elsewhere.

As well, I believe in the purpose of my work. Caves have always been a source of fascination for humanity, but their extensive exploration is somewhat recent. Underwater caves are one of the last frontiers of discovery on earth. Surprisingly, we know more about outer space than inner earth, and this is a problem. With each season that passes, the science of "water moving underground" becomes more and more relevant as we try to protect our most vital resource. My highly specialized skill set allows me to be a scientist's eyes and hands underwater, and I often work with biologists discovering new species, physicists tracking climate change, and hydrogeologists examining our finite freshwater reserves. I've found grim sources of under-water pollution, the roots of life inside Antarctic icebergs, and the skeletal remains of Maya civilizations in the cenotes, or sink-holes, of the Yucatán Peninsula. Underwater caves are museums of natural history, protecting rare life forms that teach us about evolution and survival.

In a workday, I might swim through the veins of Mother Earth, into conduits in volcanoes or cracks in monstrous bodies of ice. I dive below your homes, golf courses, and restaurants. I follow the trail of water wherever it guides me. And when the passages pinch impenetrably, the water still flows, emanating from some mysterious source. The journey is endless. It beckons me to dive deeper into what is humanly possible.

I am not fearless. I'm alive today because I've learned to embrace fear as a positive catalyst in my life. As I dwell on the threshold of darkness, I might be scared, but I don't run away. I dance in the joy of uncertainty.

IN THE BEGINNING

1967–1990

MY EARLIEST MEMORY is of almost drowning. Two years old, and out of my mother's sight, I tumbled off the dock of the cottage we rented for summer holidays and landed face down in the lake. I was too tiny to right myself, but the first wash of water had triggered an ancient instinct to hold my breath. I floated silently, nebulous sensations searing a permanent first record into my brain: the bracing water, the enticing colors, the gentle rocking of the wavelets. I drifted peacefully.

Then brand-new blue canvas Grasshopper sneakers landed in front of me with a splash and an explosion of sand, and the world slammed back into focus. I was immediately snatched from my watery haven. My mother was screaming, and I was giggling. According to family history, Mom walked on water that day.

I could have drowned, but something else happened instead. The die was cast. My life would become a quest to explore this underwater world, this place of adventure and solitude, where anyone can break the bonds imposed by gravity.

I further developed my fascination with water the following summer, on a beach in Cape Cod. It was the first time our family took a road trip outside of Canada, and our home for the week was a small beachfront strip motel. Though the beach was cool and windy, my brother, Gord, and sister, Jan, were as eager as I was to dig in the sand and swim in the ocean. I looked out over the water, but the far shoreline was out of sight. I tried to imagine what lay beyond the salty air and blue horizon. This was the first time I appreciated the expansiveness of the planet.

Our parents tried to impress upon us the brutal force of the ocean, but always experimenting, I edged closer and closer to the water. Soon, the ticklish surf was lapping at my toes. When a big wave toppled me, pulling me head over heels into deeper water, my parents both jumped in to seize me. Once again, I'd been snatched from the alluring arms of the water.

If I was going to make it to my next birthday, my parents realized I needed swimming lessons. I launched my formal swimming career at the age of four, under the auspices of the Mississauga Recreation and Parks Department, in the growing suburbs of Toronto. I loved the pool, but the changing room was torture—a vast open space with damp floors and no privacy. Tall for my age and known as "a big girl," I was already horribly self-conscious. To avoid being naked in public, I changed inside my clothes, struggling like a baby bird trying to emerge from an egg. (This skill served me well later in life. As often the sole woman on an open boat, it helps to know how to change relatively discreetly in public!)

The bathing cap I had to wear felt like a particular form of persecution. As thick as a bath mat, the foul-smelling white rubber cap snarled my long hair. Adorned with gaudy flowerets, the caps brought girls further unwanted gazes and giggles. I always wondered why the boys were exempt from this torment. It was so unfair.

I wasn't afraid of drowning. What intimidated me was the other kids. I was happiest when I was lying face down and peering into the depths of the pool. I was transfixed by the color and clarity of the water; every detail of the pool was more fascinating when viewed through this liquid lens. I wanted to touch the dimpled ceramic tiles on the bottom and the cold metal filter grate. But apparently my prone position wasn't considered normal, because at swimming-class graduation time my teachers held me back. They told my mother I likely wasn't strong enough to support my head above water, let alone swim. In actuality, I don't think I was all that interested in propelling myself. I just wanted to float.

My second attempt at swimming lessons was more successful. This time, I graduated and earned a treasured prize: a golden-yellow cloth patch in the shape of a tadpole that read "Pollywog Swimmer." I still proudly display that patch on my desk, beside my advanced level "Superfish" badge.

I STARTED MY expeditionary career before first grade. As the youngest of three kids, I was given the longest leash by my parents. My friend and neighbor Jackie Windh and I constructed backyard snow tunnels, explored the local woods, and dug holes we were sure would lead to China. We hiked barefoot on the forested land behind her family's boat-in cottage on Kahshe Lake. We tramped on the moss-covered granite, collecting snakes, fossils, and rocks. We had the freedom to explore and swim unsupervised. There were times I went home bloodied or bruised, but that didn't deter me from trying new things. I loved to ride my bicycle farther and farther out, turning around just in time to get home for dinner.

My grandfather had a lifetime membership to *National Geographic* magazine stored neatly in a closet under his basement stairs. I used to gaze through issues dating back decades. Louis Leakey, Jane Goodall, astronauts, and the men of Sealab were my heroes. Rather than playing with toy soldiers, Jackie and I played explorers. Holding my arms over my head, I'd declare, "I have conquered Everest!" We philosophized about the Bermuda Triangle and extraterrestrials, and visited the Royal Ontario Museum to look at the hieroglyphics in the Egyptian collection. We wanted to be nature and geology experts and talked about becoming astronauts, even though Canada lacked both a space program and female astronauts to look up to.

There was so much that I was interested in doing, but I was often told I was too young or inexperienced. I read ahead in class and asked for extra work from my teachers. I begged to be allowed to take the next swimming program.

At a young age, I also learned I was different from other kids. In grade two, on the dusty playground of Munden Park Public School, a bully kicked me and threw burrs in my long brown hair, creating a rat's nest that my mother would have to cut out later that night. On the first day of grade five, I found a note stuffed into the front of my workbook that read "From the class: We hate you." That day had begun carefree as I ran down the road to begin my final year of elementary school. I had skipped grade four and was being accelerated into a class filled with older kids, and I couldn't wait to meet my new friends. Instead, I was greeted with painful bullying. Girls on the cusp of adolescence have a difficult divide to cross. At ten years old, I was still a naive child while many of my older classmates were already blooming into young women. The gulf that separated us left me feeling isolated. My teacher became my best friend.

Outdoor adventures served as both my escape from social pressures and an opportunity to explore the ways in which I was different. I joined the Girl Guides, and my family created adventures too. We rarely sat around on weekends. We paddled our canoe on the Credit River and hiked our favorite routes on the mighty Bruce Trail. Following the white paint marks on the tree trunks felt like being on a scavenger hunt, but I thought there was nothing more fun than crawling through crevasses in the limestone into cool open spaces. The smell of cedar and pine still takes me back to those damp caverns filled with wet leaves that rustled under my soles.

When I was a teenager, after my sister and brother graduated from school and moved out, Dad and I sometimes took weekend canoe trips together. At dawn on a Saturday, we would drive north to Big East Lake. We'd be paddling by ten while the fog was still burning off the water. By late afternoon, we'd pull up to the edge of a pink and gray granite island covered in twisted pines. There, we'd build a roaring fire to cook steaks and potatoes soaked in red wine. On a fall day, the lake would reflect the warm red and golden tones of the sugar maples. Occasionally a turtle would pop up and disturb the glassy surface, snorting for a breath before submerging again.

I moved away from home at the age of seventeen, certain I was ready to leave the family nest. I was close to my family, but although I had yet to graduate, I never felt like I fit into the high school crowd. I was uneasy in my body, self-conscious about academic achievement, and uncomfortable with small talk. I had heard the whispers of others who teased me about my fashion choices and weight. I felt older than my classmates and was ready to either escape or forge ahead. My independence eventually won out and I moved in with two college-aged graphic designers I had met through a part-time job I had at a fabric store. Although supporting myself was tough, I was eager to test myself in the adult world. I learned to be extremely frugal, saving every scrap of leftover food and turning the heat down so low we needed to wear ski jackets in the drafty old house.

But as I edged out of my teens and into adulthood, something changed. The shift was imperceptible at first. I became more and more focused on my career, on security and stability—I started hiking less, exploring less, and working much, much more. By 1990, I was, by all appearances, successful. I had graduated from university a few years earlier and already co-owned a small, thriving advertising business. My partners and I had an enviable client list, and I found my work as a graphic designer intellectually stimulating. I lived in a beautiful apartment in the High Park neighborhood of Toronto, but I had little time to enjoy it. I began every morning with a run around Grenadier Pond, but then shackled myself to my desk before the sun even poked over the horizon. I was working sixty to eighty hours a week, designing corporate logos and coordinating advertising campaigns. I had no time for hobbies, I had to skip family events, and I was losing touch with my friends. Despite this, my family and friends applauded me for making sacrifices to achieve prosperity. But even though I was successful at work, I felt as if I was wearing clothes that didn't fit. I was drained and unfulfilled. A growing sense that my work lacked importance made me feel incomplete. That carefree child who'd lived for adventure had been cleaved from my adult self.

I was twenty-seven years old and knew I was at a crossroad. I could continue on this path. Or I could look for a chance to do something extraordinary.

SURVIVOR

1986

IT WAS A chilly night in the spring of 1986. I was an art and design student at York University, in Toronto, and I was moving into an apartment on the edge of the Lawrence Heights neighborhood. To my west was an area referred to by both police and residents as "the Jungle." It was the crime and murder capital of the city, so I wasn't really seeking out new friends when I moved my meager belongings into the story-and-a-half dwelling wedged into a minuscule lot on a busy street. The house was no beauty queen, but it was close to the subway station and, more importantly, it was really cheap.

I spent my first night in the house alone, nestled in a cozy heap of comforters on a mattress laid on the bare wooden floor. I had won the lucky draw of straws among my four roommates and selected the upstairs bedroom with a view over Lawrence Avenue West. My girlfriends and I were looking forward to our year of living off campus, away from the noise and distractions of the dorms at York. I borrowed my mom's car to move my drafting table, an old chair, and a classic staple of the starving student's decor: a bookcase made of bricks and wooden planks. My new room must have been a full apartment at some point in its past. It still contained the remnants of a kitchen with a deep stainless sink and avocado-green countertop. An accordion-type folding door kept a bit of warmth in the small room.

An old alarm clock with flip-down numbers was perched on the Labatt's Blue beer-case-nightstand next to me, and a Pink Floyd poster almost covered the water stain that ran down the sloped ceiling to the pile of cayenne pepper on the windowsill. The spicy powder was my feeble attempt to discourage the army of ants that were tossing chunks of plaster out of a hole in the wall. It was spartan, but my little bedroom already felt like home. A thrumming, regular roar from nearby subway trains oscillated in time with the traffic on the busy street below my window. It wasn't a place that a mother would love, but it was my new home, and I knew that with four athletic female roommates soon to move in, it would never be as quiet as it was this first night.

With a smile, I drifted into a relaxed sleep, dreaming about my summer job and the school term ahead. I had survived the first two competitive years of my program and had passed my portfolio review with excellence. I had earned a partial scholarship for my fine arts degree and landed a coveted job as a bartender that would help cover the costs of the following year. Life couldn't be better.

About one-thirty the silence of the empty house was shattered by a loud bang downstairs. I awoke disoriented, trying to figure out what was going on. Creaking wooden floors revealed the unmistakable presence of someone else in my new home. I had the only key, and my roommates were not due to move in for days, so who could it be? I was absolutely terrified. I couldn't even call 911 because we didn't yet have a phone connected. Perhaps the nosy landlord was coming to check up on me? I was trying to rationalize why someone would be in the house, but deep down I knew it was a burglar. Instinctively I pulled the covers high over my face and sunk deep into the blankets. I could hear the intruder downstairs, opening drawers in the kitchen and snooping through empty cupboards. Paralyzing fear left me gasping for breath. My heart was beating faster and harder than I had ever felt before. I was sweating and shaking beneath the blankets. I hoped this was a nightmare, that I would soon wake up in my peaceful, cozy nest.

Then, as quickly as I had given in to the fear, I started to think through my options. I needed to defend myself if he made his way upstairs. I slid out from under my blankets and crept to my window. But the tumble into the speeding traffic below would likely kill me. I dropped into a crouch in the corner of my room, hugging my knees and watching the numbers on my clock radio flip over, clicking one by one, while the burglar opened closets and doors downstairs. I wondered if I could squeeze my body into one of the cupboards below the sink in my room, then worried what he would do if he found me there. I pulled a brick from my makeshift bookcase, thinking I could throw it if he came upstairs. But what if he took it from me? I didn't want my weapon to become his.

1:38 a.m. He had been in the house for the longest eight minutes of my life. Dizzy with fear, I began taking deliberate, heavy steps around the room. If he heard me and knew he was not alone, perhaps he might leave. The soundlessness from downstairs only increased the pounding in my ears.

The numbers flipped over. At two o'clock the subway station and the nearest payphone would be locked for the night. I had to get out of the house soon. I pulled a thick woolen poncho over my pajamas, grabbed two slim X-Acto knives from my drafting table, and continued to tramp heavily on the floor. The very act of stamping my feet somehow gave me strength. Standing was a powerful posture that seemed to erase some of the helplessness I felt. My poncho smelled of canoe trips and campfires, times of comfort and safety with friends. I wished it could magically protect me now.

Finally a sound: the tiniest creak on the stairs. 1:43. I raced over to the drafting table and crouched underneath it, clenching my weapons. The pointed silver blades would be an extension of my hands, razor-sharp and ready. I was sweating profusely, and could not stop my body from shaking. Even my teeth were chattering.

I meditated on my strength while staring at the numbers on the clock that teased and flipped over again. Each roll of a minute was bringing me closer to the ultimate confrontation of my life.

1:48. I heard a muted click when the dim hall light illuminated the first glimpse of the intruder. I saw the shadow of his feet below the thin folding door and the crest of his curly hair backlit through the gap at the top. A delicate barrier separated us. Predator and prey. He stepped into the hallway closet outside my bedroom door. The sound of the metal hangers scratching along the rod chilled me like fingernails on a chalkboard. He was searching through my clothes. He was building a picture of who I was, just as I was imagining what he looked like.

1:52. As he stepped out of the hall closet a rush of electricity coursed through me. I was completely present, completely alive. I watched his large shadow pass by my door again, and then I heard him in the bathroom. I heard the medicine cabinet open—items picked up and set down again. My hairbrush sounded like shattering crystal as he thrummed his fingers across the plastic bristles. Then the sound of porcelain scraping against porcelain. He was pulling the lid off the toilet tank. Was he going to beat me with it?

1:55. The subway station a block away would close in five minutes. I had to get there so I could call for help. But then his feet were back at the base of my door. The handle jiggled. I uncurled my arms from around my knees and stood up behind the drafting table, completely vulnerable and exposed. I placed my hand on the desk lamp. I wanted to hide and cry.

I could hear him breathing.

Hold the knives. Don't let go of the knives.

I smelled him through the thin barrier of the door. A blend of sweat, street, and rage.

Hold the knives. Don't let go of the knives.

There comes a moment when you're confronted by fear when you have to make a choice. Surrender, or fight. And in that moment

of clarity, we find what we're made of. I took a deep breath and
steeled myself. I realized that I was capable of anything. I would
survive. Whatever it took.

1:58. He sucked in a deep breath and launched through the folding door, pulling it off its track. Even though I had tried to prepare
for this moment, I almost jumped out of my skin. I turned on the
desk lamp and pointed it at his blanched, perspiring face.

"Who are you? Identify yourself!" I yelled.

He smirked, his eyes red-rimmed and crazed, and then lunged
toward me.

I leapt forward, and with every bit of strength I could find, I
slashed at him with the knives. With my right hand, I sliced him
from his right shoulder to his waist. The friction of flesh didn't
stop me. I thrust the second knife downward across his chest. He
stumbled back and looked down. A crimson stream dribbled into
his cupped hands. He chuckled maniacally. He looked up at me,
right in the eyes. Then he clumsily turned on his heels, walked
slowly back through the broken bedroom doorway, and retreated
down the stairs.

No matter how hard I gulped, I couldn't get enough air. I slumped
to the floor, fighting to breathe, still holding the knives.

I was overwhelmed by relief, but I knew the danger wasn't over.
I forced myself up and crept out of the bedroom. I looked in the
closet. I thrust open the bathroom door and kicked the shower
curtain hard enough to spring the tension rod from the wall. I ran
down the stairs at full speed, through the empty living room and
out the open front door. I didn't stop running until my nose was
pressed against the glass of the subway station door a block away.
Tears ran down my face as I pounded on the glass. Wheezing to
get the words out, I begged the station worker to let me in. I must
have looked desperate enough that he reluctantly opened the door
and offered me a chair. Someone called the police while three
brawny transit staff tried to pry the knives from my trembling

fingers. Only once the police arrived, and my safety felt ensured, did I drop my two bloody X-Acto knives to the floor. It was over.

AFTER DAYBREAK, THE police dropped me off at the university. I went straight to work, opening up the college pub I managed part-time and putting on the first of many pots of coffee. The police must have taken me home to get clothes, because I was no longer in my pajamas and wool poncho. I have no memory of that time.

I lay awake for three full nights before exhaustion won. I was still shaking. Someone brought me a baseball bat for future home defense. It was a joke, but I clutched it for hours, then kept it at my bedside for years.

Over the next few weeks, the fear bubbled to the surface whenever it was dark. When I managed to sleep, every sound woke me up in a cold sweat, so sometimes I avoided sleeping altogether. Even in the daytime, I couldn't shake visions of my attacker. I searched crowds for his face, afraid I might see those vacant, soulless eyes again. I tried rearranging my bedroom so it felt like a different place. I kept my back to the wall in every room.

A few months after the attack, over a beer, my roommate Kim told me I looked terrible. I told her that constant nightmares were robbing me of sleep. I was expecting a patient and understanding response from her, but instead, she looked at me and said, "Get over it!"

I was offended. How could she possibly understand the terror I had experienced? She hadn't faced a burglar, fearing she would be raped or killed.

"You can't go through life cowering in corners, Jill," she continued. "You're letting him hold you hostage."

I knew she was right. The enemy wasn't still lurking in my house; rather, it was inside me. I had been running away from the things that scared me to protect myself, but in reality, I was becoming feebler. Living in a hypervigilant state was exhausting, and it was hurting me.

I had a choice to make: I could continue to be a victim or I could try to rise above the experience. I couldn't change that the attack had happened, but I could change what happened next. I was coming to the same conclusion I'd had when the burglar was in my house: once again, I had to believe in my own power.

SIX MONTHS LATER, I was visiting my parents in Mississauga. Coming home from university was always special. Mom cooked my favorite meals and showered me with attention. After I moved out, she had quickly converted my childhood bedroom into an office and craft space, but that didn't bother me; I was just as happy to sleep in my brother's old room.

That night I fell asleep with a full belly, safe and untroubled in the comfort of home, my parents just one room away. I hadn't told my mom and dad about the burglary because I didn't want them to worry. Ever since my conversation with Kim, I'd been managing my anxiety much better, though I sometimes still had flashbacks. But my parents' neighborhood was a safe place—it was a sheltered enclave where kids roamed free and people looked out for each other. The homes faced inward to a court, which created a real sense of community. There was a Crime Watch with nothing to watch, and a TIPS Hotline that never rang.

Sometime after midnight, I woke up from a nightmare in a pool of sweat, my covers on the floor. I could hear my parents stirring beyond the wall, so I forced my eyes closed and tried to slow my hammering heart.

Then I heard a faint rustling downstairs. With unbearable dread, I was certain that someone was in the house. A second later, the security alarm erupted.

I leapt out of bed, choosing once again to fight. My brother had been given a set of steak knives for Christmas, and he'd left them on the dresser. I thought three in each hand should do the trick.

I ran into the hall, cursing at the intruder. I met my dad on the landing. "Who the hell is here?" I yelled down the stairs. Dad looked at me in shock, and then bounded down the stairs ahead of me. He disabled the alarm and looked at me sternly.

"What's gotten into you?" he asked. Mom had been so sure it was a false alarm she hadn't even bothered to get out of bed. But I knew what I'd heard.

I raced from room to room, desperate to find the intruder. My father must have thought I was completely overreacting. With a gentle voice, he calmed me down, and together we checked every door to try to figure out what had triggered the alarm. He convinced me no one else was in the house and that the alarm had just malfunctioned. Satisfied, we went back to bed.

Eight minutes later, the quiet was interrupted again by the clanging alarm. The hiding burglar had tripped a sensor and was still in our home.

This time, I raced straight downstairs ahead of Dad, determined to get to the thief. At the bottom of the stairs I smelled sulfur—the intruder must have lit a match in order to see. The sliding-glass kitchen door was ajar, and I watched a shadowy figure jump over our backyard fence and into the neighbor's yard. I stepped out onto the porch but couldn't chase him. My courage had boundaries, and that night, it ended at the edge of the patio. The figure, silhouetted in the moonlight, bounded over the neighbor's gate to get to the street, but, missing his mark, he landed on his belly on the gate's center spike. He managed to extricate himself and then I watched him escape into the night.

Two break-ins in the same year were inconceivable and terrifying, but they were a gift. Nobody could have taught me how to access my strength when I needed it the most, but these new experiences did. The second break-in gave me a chance to follow a different script, to bypass most of the vulnerability and victimization I'd felt during the first intrusion. I learned to temporarily depress my fear and instead jump to action. Slow my beating heart. Breathe away the stresses that wouldn't serve me. This time, I'd refused to surrender to panic. And I did not want to give up my power to another human being—or to fear—again. I could take the next pragmatic step toward survival.

For the rest of my life, I vowed to practice the skills needed to be a survivor. I would not hide under the covers again. I would face challenges with fierce will and optimism.

Four years later, exhausted and overworked, I chose to face the fear of disappointing others. To truly thrive, I knew I needed a complete reset of my career. I decided to take a plunge.

AN ALLURING MISTRESS

1988

I PULLED THE dump valve on my buoyancy control device, causing a whoosh of escaping gas, and started sinking into a new world.
I felt as though every cell in my body was vibrating with excitement. Although I was just completing my initial scuba certification, I sensed I was on the cusp of a new beginning. I had wanted
to learn to dive since the first time I saw Jacques Cousteau on TV
as a young girl. Now I was making it happen and I was ecstatic.
It was a world apart from my drafting table at work, where I felt
myself shrinking. Underwater, in this hidden world, I was expanding with the potential to do something truly remarkable.

Muffled sounds were coming from every direction and time
seemed to slow down. I deeply exhaled and surrendered into the
abyss. Some of my classmates were flailing as they descended, eyes
wide with fear and resistance. I felt a tingling surge of adrenaline
that heightened my senses and triggered a shiver through my
spine. The blue void around me drew out my imagination to consider all the things I might experience in my future as a diver.
Would I one day swim with whales or sharks? Would I find trea-
sure in the bowels of a shipwreck?

I dropped through a shimmering mixing zone of water, where
the temperature plunged to 37°F, but the chill faded quickly.
Moments later I was thirty-two feet deep, hovering above the

boulder-strewn bottom of Lake Huron. Smoothed by the action of repeatedly rocking waves, the rounded stones sounded like bowling balls on a return ramp, clapping and rumbling with each passing breaker. A gentle ebb from two-foot swells pitched us back a few feet after every lunge forward. I gave in to the sway of the oscillating water and copied the moves of my instructor, Heather, who was swimming with ease toward a dark limestone archway. As I peered into the blackness ahead, my heart rate spiked. The darkness was calling me forward to something absolutely thrilling. It was both a literal and a figurative gateway, and I would cross its threshold to a new chapter in my life. Three feet forward, two feet back in the surge . . . it seemed to take an eternity to breach the entryway and swim into this place we called the Cave.

This first dive into what's called an overhead environment—one that doesn't have direct access to the surface—took place not far from the terminus of the Bruce Trail on the same sunny weekend I was certified as a new diver. It was rather advanced for an initiation dive, and left a lasting impression in my mind. After three successful dives on sunken shipwrecks in Tobermory, Ontario, to demonstrate our skills as new divers, Heather was rewarding us with a special treat. Straddling the orange pontoons of an inflatable Zodiac boat, we tightly grasped the thick nylon ropes like cowboys on broncos. For an hour we sped across the surface of the lake toward the dive site, along the edge of a beautiful rock face that was blazing with the vermilion hues of late-afternoon light. Protected from head to toe in black neoprene wetsuits, we welcomed every trumpet of spray that blasted across our faces. The anticipation made my body prickle and my face flush. The fourth dive of my life would lead to a proud moment when I would receive a card that certified me as an Open Water Diver.

Now, passing over the lip of darkness into the gaping maw of the cavern, I was momentarily disoriented. Beneath the rock ceiling, a single turquoise light ray sliced diagonally through the water ahead of me. Like a tractor beam, it pulled me forward and upward to float in the flickering light of a large open room.

Then the water in front of me exploded into a fury of white bubbles. Clambering legs poked out from the burst, and a scrawny white figure in an orange Speedo kicked quickly toward the surface. I realized I was inside a grotto with a glassy mirrored surface inside an echoing chamber of rock.

I later learned that this was a favorite destination for thrill-seeking hikers who tramped down a forested path to reach this spot. They would crawl into the cave through a small skylight and a larger opening—a doorway, we called it—along the edge of a cliff and take their place in an orderly lineup for a fourteen-foot leap into the water. The daylight I had seen from below was streaming from these portals, illuminating the depths of the cave like the sun through a cathedral's clerestory window. The light danced rainbows on the sandy bottom. It reminded me of that first plunge off the dock at the cottage. The only thing missing this time was a pair of navy blue sneakers and someone pulling me back.

With a single breath, I could move in three dimensions: inhale and rise to the ceiling or exhale and sink in slow motion. It was as though I had the superpower of levitation. I transcended gravity. I was no longer a big girl or a clumsy earthling who needed to heft her weight up the rock wall of a cave. I was flying free in the water.

This innocent little jaunt was a far cry from real cave diving— I was never more than a few feet from a safe exit—but it whet my appetite and stimulated my imagination. Somehow, I knew then that diving would be something I'd do for the rest of my life, and after that moment, every time I felt trapped by the demands of my job, I would take a deep breath and close my eyes to be transported back to the grotto, to float in the sapphire beam of sunlight. Each time I went diving, the call to the water was louder. The satisfaction of a successful job and hefty paycheck were no longer enough to fill my heart.

At the time, I perceived only the magic of my first diving experiences and none of the risks. I had neglected to consider the danger imposed by a rock ceiling and, at the time, I had no idea how lucky I had been. Other divers who venture into overhead environments for the first time have not been so fortunate. Underwater caves offer up a deceptively easy way to die, and without training, what you don't know might easily kill you. Training as an open-water diver teaches you that in an emergency you can always swim to the surface, but in a cave or other overhead environment, that is not possible. If you are not prepared to deal with a bad situation such as malfunctioning equipment or disorientation, then you can die in the blackness of a water-filled room inside a cave.

The 2018 incident involving the Wild Boars football team trapped in Thailand's Tham Luang cave brought the risks of cave diving—poor visibility, fast water, getting lost, and running out of air—into the public eye. Retired Thai navy diver Saman Kunam ran out of gas and drowned in the overhead environment simply because he failed to watch his air gauge. When the team of civilian cave divers took over the rescue efforts, a close friend of mine remarked, "You'd have to knock me out and tie me up to get me into an underwater cave." And that is exactly what had to be done in Thailand. Each young boy had to be bound to a rescue sled and anesthetized so he could be hauled to safety through the submerged tunnels. Even one of the skilled British rescue divers, Chris Jewell, had a close call when he let go of the safety line while struggling with a rescue sled holding a sedated child. For Jewell, it was four minutes of terror while he groped in the dark to reorient himself and find a way to the next air pocket.

Soon after my open-water class, I learned about the risks of cave diving when I picked up a copy of a primer called "Basic Cave Diving: A Blueprint for Survival." The little blue book was written by a pioneering explorer with one of the coolest names I had ever heard. Sheck Exley didn't waste any money on graphic design; the stapled booklet appeared to have been photocopied from a type-written manuscript. But the information inside was lifesaving.

Exley, a high school math teacher from Live Oak, Florida, was fixated on statistics and accident analysis. Having had numerous brushes with death in his cave-diving exploits in the late 1970s and '80s, he decided to take a statistical look at why people were dying at such an alarming rate in Florida caves. He himself had recovered the bodies of divers who had become lost or run short on breathing gas. Exley wanted to share lessons from those tragedies with the diving public. At the time, there was no formal training for cave diving, so Exley wanted to educate divers to become safe cave divers. His safety rules would have to become my touchstone if I intended to stay off Exley's lists. If I wanted to survive my future cave dives I would need to learn more about gas turnaround pressures, running guide lines, and proper equipment redundancy—all things that were completely foreign to me then.

So although my first peek inside the grotto in Tobermory, with an instructor close by, empowered me with a false sense of security, within a year I'd develop a healthy respect for Mother Nature's fury and an understanding that it *could* happen to me.

ONE OF MY first diving trips after being scuba certified included a perilous crossing to Canada's far western islands over very unpredictable seas. Our journey began in Port Hardy, on the mist-shrouded northern tip of Vancouver Island. Colorful wooden shacks whispered in and out of focus through meandering banks of fog. Spruce trees the height of apartment buildings loomed over gravel roads running through the rainy forest. In gaudy hooded rain jackets and hiking boots, eight dive club friends and I scampered along a broken cantilevered dock to a weather-beaten sixty-five-foot boat. The *Clavella* was dwarfed by the bigger oceangoing vessels in the crowded port. Her dark wooden decks offered a vivid contrast against the white hull decorated with Cape

Cod–blue trim. Most of the items on the minimal deck space were tied to the vessel with tattered old rope, evidence of the challenges of sea travel.

It all should have served as a warning. But at that moment, I was completely focused on a total break from work. This wasn't just a vacation; it was an adventure. For the next two weeks I would be off the grid and in the zone. Nobody from work would be able to reach me, and I could focus on living the kind of life I envisioned for my future—outdoors, action, and creativity. I felt as free as the powerful bald eagles that were soaring over my head.

We were greeted at the dock by John deBoeck, an affable, strong man with a gentle voice. His skin was etched by his many years at sea, and crow's feet were permanently engraved by a constant grin. His mildewy white tennis shoes could have done with new laces years ago, and the cuffs of his battered sweatshirt were as worn as his freckled, meaty hands. He welcomed us on board and showed us to our stacked bunks below deck. Apologizing for damp mattresses, he explained that his old vessel "sometimes springs a few leaks."

We were given an orientation to the safety devices and procedures of the boat. "The bamboo pole and plywood below deck can be used to seal a blown porthole," he said matter-of-factly. "Always tie the pots to the stove, and never leave the seacocks open on the head. If you have to go on deck, make sure that a buddy is aware, and if you must be sick at night, tie yourself to the rail."

"That doesn't sound like the travel literature!" I said, hoping that he had not overstated the vibrant soft coral walls, leaping orcas, and relaxed cruising through fjords framed by spruce forests. Our plan for the trip was to motor up British Columbia's Inside Passage and then cross over to a remote island group off the Pacific coast, Haida Gwaii, formerly known as the Queen Charlotte Islands and the sacred homeland of the indigenous Haida. I had a new Nikonos underwater camera, and I was looking forward to photographing a marine paradise filled with whales, Pacific white-sided dolphins, giant octopuses, and other fantastic creatures of the emerald seas. But first, we had to make a twenty-four-hour crossing of Hecate Strait in heavy seas and ripping tidal currents. The isolated islands were days from help in cold northern waters. John insisted that *Clavella* was up for it, but I wasn't as certain.

After four days of incredible diving, we set out to make our crossing to Haida Gwaii. Now accustomed to the turbulent tidal currents, I had gained confidence in Captain John's skill at picking us up out of the towering seas with a twenty-eight-foot-long skiff that was towed behind *Clavella*. I was excited to finally head over to the tempestuous but undiscovered waters of Haida Gwaii. It felt like real exploration.

But as we left the protected coves of the Inside Passage, things changed quickly. A capricious storm arrived, battering us with waves that grew as night came on. The waves relentlessly pounded against *Clavella*, shaking it with deafening wallops that felt like they were going to break us apart. Captain John wrestled the big wooden wheel while the decks were washed and drained with every rise and fall of the sea. As one of the few people who was not retching violently, I stumbled to the back of the vessel to check on our diving skiff, which was chock-full of our scuba gear. It was strung to the stern on a two-inch-thick line and was bucking like a wild bull. I was instantly soaked with freezing water that crashed over the rails.

In the starless night, I quickly moved back from the stern, through the salon toward the wheelhouse. Then, without warning, a rogue wave slammed the boat broadside and threw me to the floor. The vessel's navigation systems—radar, loran, and communications gear on the top of the wheelhouse—were stripped away by the wave. While many of my fellow passengers and crew started climbing into dry suits for survival, I watched in horror as the skiff rolled over. With a shuddering groan it slowed the boat down. Moments later, Captain John decisively severed the towline with a swipe of his knife, and I watched my first set of scuba gear sink to the bottom of the ocean.

I could have been angry at the loss of my uninsured equipment, but instead I was weirdly gleeful. I could have been dead on the bottom, in 660 feet of water, but by dawn, John's skillful seamanship with a hand bearing compass had delivered us to a sheltered cove. We had survived the challenges of the dark and ferocious sea. For me, it was an early lesson in gratitude and smart decision making. Our lives were far more valuable than "stuff." More importantly, I gained a new respect for the temperamental nature of the ocean. I understood how quickly conditions could turn deadly.

It would have been easy to quit diving after such a brush with disaster. Nobody would have questioned my decision. After all, how many would want to replace $3000 worth of equipment? But the adventure did not deter me. Instead, it further solidified my desire to learn new diving skills and increase my comfort in and on the water. I bought better, more advanced dive gear. That summer, I enrolled in every continuing education class that was posted on the schedule at The Diving Store in the west end of Toronto. Rescue Diver, Wreck Diver, Underwater Photography, and Divemaster programs soon followed, and even when the snow started to fall, I kept it up, learning ice diving, deep diving, and anything a mentor or instructor would offer. With new friends from the scuba shop, I made weekend excursions back to Tobermory and to the St. Lawrence Seaway at the eastern terminus of the Great Lakes. I volunteered to help student divers on evenings and weekends even while my graphic design job was still eating up sixty to eighty hours a week. The damp carpeting in the trunk of my little red Nissan started to smell like mold mixed with chlorine. I ate my meals in the car or at my desk, not wanting to waste precious time when I could be diving or reading books like *The Darkness Beckons* or Cousteau's *The Silent World*.

After earning my scuba instructor credentials at the Wet Shop, in Toronto, I slipped away from looming work deadlines to teach scuba classes in the evening. On Friday afternoons, I escaped as early as possible to beat rush hour and make a pilgrimage northward to go diving. My earnings as a scuba instructor didn't even cover the gas in my car, but I wasn't counting pennies in those days. Although I loved my creative work, doing it inside the repressive four walls of an office wasn't cutting it. While I sketched out concepts for Canon Canada ads, I was thinking about where I would go for my next dive trip. I had a running wish list of scuba gear I needed to buy and dive charters I wanted to book. I yearned to live outside the confines of an office. I was happy running in the park near my apartment. I was joyful on a boat on the lake. But my claustrophobic office was stealing my soul.

My partners' ideas of success did not resonate with me. They lived for their children and three-bedroom suburban homes, but I knew in my heart that those traditional landmarks were not in my future. I dreamt about building a creative life in the underwater world. I wanted the life of the photographers I saw featured in *Skin Diver* magazine and *National Geographic*. I wanted to travel, to explore and document things that had never seen the light of the sun. I wanted to share the beauty of the underwater world with other people, and shooting photographs and films might be a way to support myself while exploring caves and other unseen corners of the earth.

I'd be lying if I said it was a great, daring leap. It was more of a gradual metamorphosis; a butterfly emerging from a cocoon to fly free. Decisions of such considerable magnitude are hard to make, and they take time. Family expectations and convention forced me into a convenient and predictable life-script: Grow up. Go to university. Work in a professional career. Have kids. Toil until retirement and delay satisfaction until you have earned it. But what if I never made it to retirement? Why couldn't I live happily now and let the rest of my life sort itself out? I felt as though my thoughts were traitorous or selfish, as if I was betraying society, my family, or womanhood. Each diving experience shifted me toward a road less taken, but I wondered if I would lose my friends and family if I abandoned the expected path.

The inertia was tough to overcome. I had become accustomed to substantial paychecks and a comfortable lifestyle. My apartment was right in the heart of a hip neighborhood and on the edge of the closest thing to wilderness in Toronto's inner city. I could enjoy good restaurants, buy the best dive gear, and still put away money for vacations. Trading all of that for the meager subsistence of a diving instructor seemed reckless, and the weight of disappointing my father was paralyzing me. I had borrowed money from him to launch my business, and although I had already paid him back, I worried that he would be saddened if I walked away from his concept of success. I would need to develop skills and connections to move my underwater career forward. I still had a long way to go. But after much agonizing, I decided to move to the Caribbean to chase my new dream.

Predictably, my friends and colleagues were shocked. When I told one former university professor that I was leaving the country to become a professional diver, he was so uncomfortable that he cut our phone conversation short. Like my business partners, most of my friends were working hard toward traditional goals. Some were starting families and others were climbing the career ladder. When we had graduated university, a very close friend challenged, "Race you to the top." When I told her I was leaving my business, she said, "But, Jill, you have so much potential."

It should have been easy to walk away, but choosing an unconventional life is worrying, no matter how passionate you feel about it. Still, with great trepidation, I struck a deal with my partners to buy me out, paid all my debts, gave up my apartment, and bought a ticket to the Cayman Islands, my new proving ground. Dissolving the business partnership left me no option other than to make a success of myself, but at least I could count on a safety net of monthly payments from them. I finally had a chance to pursue my dreams.

DIVE RESORTS ARE happy places, because everyone comes for what they hope will be the best weeks of their lives, and I was no different. As soon as I arrived at the modest twelve-room divers' motel, I knew I had made the right choice. "This is awesome!" I declared, throwing my arms around my new boss and friend, Danny Jetmore. As I inhaled the sweet fragrance of pink frangipani blossoms, my optimism grew. Stripped of everything but a few personal belongings, I felt unencumbered, free to succeed. I made quick friends with the other staff members and they soon felt like family. I thrived in the tropical warmth, running, biking, and swimming my way toward fitness as a triathlete. I celebrated shooting each precious frame of slide film, visualizing the images I would later pick up from the camera store. I scoped out my shots, considered the angles of light, then waited for precisely the right time of day when everything would be perfectly illuminated. I dived, snorkeled, and swam every day and I got to know the inhabitants of our local reefs like best-loved pets. I felt, finally, that I was in the right place.

A few months into my new job at the Cayman Diving Lodge, I was saddled with a big test that made me reflect on my decision to move away from Canada. I was sitting at the open-air bar in a warm evening breeze when the phone rang. I hadn't heard from Rick, a friend and former colleague, since I left Toronto, so it was a surprise to hear his voice on the other end of the crackling line.

"Jill, I went to your old office to see about getting some work done. Your sign is gone, and there's a note on the door saying the company has closed."

"What? How can that be?" I asked with a trembling voice. My former partners hadn't said anything. "I haven't gotten a monthly payment in over seven weeks, but I was blaming it on the post office."

Despite a legal contract, my former business colleagues stopped the payments to buy me out. They shut down the company and immediately reopened under a new name, and I lost my hard-earned nest egg. I felt a mix of anger, terror, and bile rising from my stomach. I could subsist on my diving pay, but my entire life savings were locked up in the equipment of my old business— computers and gear that my ex-partners had now tied up in loans to fund their new company. I had just given up a lucrative career. Was my dream of being an underwater photographer about to fizzle out for lack of funds? At this point I was having fun, but I still wasn't certain I could turn this new life into a real job. I wasn't even certain that I had the talent.

I needed to clear my mind and decide what to do. I carefully prepped my Nikonos V underwater camera with my favorite wide-angle lens. I loaded thirty-six frames of color slide film and chuckled darkly about how I might not have the money to develop the film. I rolled off the stern of the forty-two-foot Newton into a flat calm sea on Grand Cayman's East End. The ocean greeted me with the comforting embrace of 82°F cobalt-blue water. I drifted down to the crest of the wall where large purple sea fans undulated in the surge. A school of dog snappers zoomed past me as a southern stingray puffed out of the sand flat beside the coral. I dropped deeper, feeling the dangerous narcosis—a confusion caused by pressure that increases with every foot of depth—softening the edges of my thinking. Deeper, past giant orange sponges, the intensifying numbness made everything feel okay.

I paused at 167 feet, the darkness beckoning me deeper. My choices were here. An alluring siren was calling me down to the infinite depths, but as I looked upward, I could see with crystalline clarity the name on the stern of the boat. It seemed as though I was perched on a precipice. I could give up and float endlessly in the caress of the ocean or choose to swim back to the challenges of the surface world. I took photos of the waving fronds of soft coral infused with dancing schools of butterfly fish. It was a stunning scene, and the rapture of nitrogen narcosis made it appear even more remarkable. Then my beeping wrist computer awakened me to the fact that I needed to make a slow ascent to help my body readjust to the surface pressure. This penalty, called decompression, could not be ignored without risking injury—or, worse, death. My air supply was dropping fast. It was beautiful there, but I had to head back to the surface immediately.

I was swimming up the colorful wall of tangerine sponges and purple sea fans when a sudden thud startled me. A critical seal on my camera had imploded from the pressure at this depth and was now spewing bubbles. That meant only one thing: destruction of the one thing I owned that could fully enable my career goals. My expensive camera and lens were now flooded with corrosive salt water. What else could possibly go wrong? Perhaps I was fooling myself, thinking that underwater photography could be a viable career for me.

That evening, I swung in the beach hammock with a strong gin and tonic and thought through my options. With my creative tool destroyed, I was forced to dig even deeper into my motivations and goals. Perhaps I needed to accept the normalcy of a daily commute and nine-to-five job. And I questioned whether I was worthy of the money I was owed, and ultimately worthy of happiness. My father recommended a lawyer to me, but I felt so defeated that I wasn't certain whether I was a victim of fraud or of my own overblown view of myself. Maybe I just needed to chalk it up to an advanced education in life.

Some say that you have to reach rock bottom before you can see clearly. After a few days of thinking, I eventually got some perspective. I was penniless, yes, and living in a room in a diving lodge, but I wasn't destitute. I had a job, access to a vehicle, and a tropical paradise all around me—hardly destitute. I made a trip back to Toronto to consult with a lawyer and got some of the best advice of my life. The elderly barrister across the desk locked eyes with me. "The question is not whether you will win this case," he said. "It's a matter of what you want out of life. I can win this case and I will get paid well, but all you will get is justice. There simply isn't enough money on the table. This will take years to run its course. You are young and bright. Think hard about how you want the next years of your life to proceed."

I had already given up a high salary, so why not sacrifice every tangible asset to achieve something intangible but critical to my happiness? I realized that I had to let go of any thoughts of revenge and commit fully to my diving career. So I returned to the island with strength and resolve instead of defeat. I had no choice but to succeed.

But first I had to find a way to save for a new camera. Then, just as I was feeling hopeless, our lodge chef, Linc, asked me to cover for him for a few days. A close family member had died, and he needed to head to Atlanta immediately for a funeral.

Linc was the lodge's professionally trained culinary genius. He kept our kitchen stocked and ensured that everyone was well nourished with food they would later rave about. Covering for him seemed impossible with my schedule of dive instruction, boat charters, and other tasks. I would need to cook for up to twenty-five lodge guests and staff. But perhaps the extra responsibilities would pay for a new camera.

"Of course I will," I said.

So I became an accidental chef. With no training other than the joy of cooking, I relied on Linc's recipes and a lot of help from the rest of the staff, and set about cooking meals in between boat trips to colorful walls and breathtaking reefs. It would have been manageable for a few days, but time stretched on. After more than a month of absence, Linc officially resigned.

We never managed to hire a new chef. Work permits were hard to come by, and no local people were interested in the position. So this was my new reality. Although I managed to share some of the tasks with other dive staff, for the next two years, my hands were full. After a few culinary disasters and a steep learning curve, I learned to make jerk chicken, chocolate mousse, and a bizarre mistake of a meal that I called Twice-Washed Curry. I figured out how to keep the pantry full with weekly deliveries and got a fast lesson in food storage and kitchen hygiene. And through it all I stayed focused on my dream career of capturing art in my underwater paradise.

Because I was saving all my wages and tips for a new camera, I turned to activities that were free. Instead of joining other divemasters and instructors at nighttime parties in George Town, I explored the wooded wilderness near the resort, climbing into holes in the limestone and searching for caves. I was taken back to my younger days hiking the Bruce Trail. I wiggled through spaces in the rock, finding small caverns and a few voids filled with trash. For some people, these openings in the ground were evidently nothing more than a place to discard refuse, but for me they were doorways to possibilities. Each dark nook offered a chance to find some new inner sanctum. I wasn't just exploring the pocked landscape; I was also exploring conceptual barriers. By passing through the threshold of darkness, I was discovering my psychological limits and potential. Each time my eyes adjusted to the dim light, I would find new strength, and with courage, I could go further.

Out in the ocean, I guided visitors through long, sinuous tunnels in the reef that were filled with fish, but I dearly wanted to find an undiscovered inland cave, a secret place to explore on my own. I wanted my own personal hideaway where I could survey, draw maps, and shoot photos. I was dancing around the edges of being an explorer and wanted to show some tangible results that supported what I already felt in my heart. Like everything I ever did in life, I enthusiastically jumped in with both feet and began a search.

Danny Jetmore, the manager at the Cayman Diving Lodge, was a little worried about my solo excursions searching for caves. He wanted me to dive with someone with more experience. "What happens when you don't come back at sunset?" he'd ask me. "Where would I even begin to look?"

"I'm very careful, Danny," I reassured him. "Nothing is going to happen to me."

That was when he introduced me to his close friend and cave-diving instructor Paul Heinerth. Danny had known Paul for more than a decade and invited him to come out to Cayman for a free vacation. I could not help but note his mischievous smile and confident swagger. His rumpled casual clothes and worn Birkenstock sandals projected a relaxed confidence that I found appealing. Watching Paul glide through the water was like observing the elegance of a manta ray. He moved with effortless ease as if the ocean was his native element. Here was someone who could teach me to cave dive—properly. It seemed as though the universe was delivering him to me.

Paul became a frequent visitor to the island, often bringing a new girlfriend with him. The morning after one of his visits, Danny whispered in my ear that Paul was infatuated with me. I had thought the attraction was one-sided. Although we enjoyed talking about diving, he always had a girlfriend with him when he traveled, and I had no reason to believe he thought of me as anything more than dive staff. Yet it turned out his frequent returns were a shy overture of interest. So I invited him to explore a cave I had recently discovered. I confessed that I had no formal training in cave diving and knew just enough to get myself into trouble. Would he help?

He eagerly returned to the island a few weeks later to explore the cave with me, traveling with cave-diving equipment, ready to join me on my day off. But again he brought his girlfriend with him, and so I assumed that the rumors of his romantic interest in me were unfounded. While his partner sipped bikini cocktails in the beachside hammock, Paul and I dressed down in jungle-stomping clothes and solid trekking boots. I was disappointed that he had brought her, because I was attracted to him, to his sense of wonderment and appreciation of the natural world. He had charmed me with his coy compliments, and I could never get enough of hearing about his early diving exploits, finding new caves and traveling to exotic locations with people like Sheck Exley. I was looking for more than a diving partner.

A few weeks earlier, residents of East End had tipped me off to a small pond where livestock watered in the bush. Drinking-water ponds are often fed by groundwater springs, but this spot had even more going for it. A local bartender told me about placing a turtle in the hole and finding the same turtle later in the sea. Residents of East End were certain that the inland pond connected to the ocean via some underground passage. Hoping there was some shred of truth to this, I spent days chasing cows in the heat, only to find they were running away from me and not toward some mythical water hole. On the sixth day out, after tireless, patient tracking, I was finally rewarded when a cow came to a standstill at a muddy pond under a sweet-smelling poinciana tree.

While my large bovine guide cooled its ankles in the shallows, I pulled two long freediving fins and a mask from my backpack. I tossed my T-shirt and shorts on a nubby tree root and slipped on neoprene booties to protect my feet. The mosquitoes immediately bit into my exposed skin, hurrying me into the water. My feet sunk into the gooey bottom of twigs and mud and one boot was almost sucked off my foot. I quickly launched myself into the stinky swamp water that was covered with sticks and leaves. As I submerged, I could see nothing through my dive mask other than brown murk, but I felt a coolness shivering up my legs. This was a sign. The cold water had to be coming from some deeper source.

It took a moment to balance my excitement and fear before making the decision to see what lurked below my feet. I calmed myself and filled my lungs from my belly to my shoulders, packing in as much air as possible. I swept aside the floating debris and pivoted at the waist, flinging my feet over my head. Two kicks later, I passed through the fetid, warm water and broke into a layer of cold. At eleven feet I noted an opening in the rock but continued deeper, wiggling my jaw to adjust my ears to the rising pressure. I shone a small light on my wrist-mounted dive computer. At thirty feet, I grabbed on to a spire that hung like a fang from the ceiling of an open space. I could barely illuminate the outlines of the tunnel that dropped downward into blackness.

The thought of going somewhere that nobody had ever seen before sent a stream of adrenaline rushing to my fingertips and toes, bringing with it a thrilling hyperalertness. I tried to absorb as much information as possible, memorizing every detail of the cave. The walls looked like they were sculpted in undulating curves interrupted by hanging columns of rock that dripped from the ceiling. I waited in the portal as long as I could until my chest started twitching with the spasms that warned me it was time to turn back. I let go of the talon of rock and relaxed my muscles, allowing myself to be lifted toward the surface. I broke through the foul-smelling water, took a gasp of air, and screamed out, "Yes!" I had discovered my first underwater cave.

Now, weeks later, Paul and I stumbled over rocky nubs and cow patties along a narrow path beaten down by thirsty cattle. He was twelve years my senior but had no trouble lugging the heavy diving equipment on his broad shoulders. Paul was a stocky, barrel-chested college wrestler, and his strength came from years of handling heavy gear at the scuba shop he had operated since graduating from the University of Florida.

His sweaty face could not hide his utter disappointment when he finally saw the small boggy pool. "Jill, the water is crap," he said, in his slight French-Canadian accent. "If there were a cave with flowing water down there, it would be a little prettier than this."

Looking at the swampy, excrement-ringed pond, I understood his doubt, but I also knew what I'd seen. "I know it looks bad, Paul, but you have to trust me. I saw calcite formations and a cave."

He took a deep breath, then smiled and offered to run the safety guide line into the cave while I followed from behind.

As with my first time here, getting into the water neatly was not an option. We waded through the oozing mud in our heavy scuba gear, further stirring the pool into a thick soup. I knelt down and flopped face first, making a less than graceful start to our dive. The sticky clay mud streamed off my legs and trailed behind me. I turned on my dive light and could see only a faint orange halo just beneath the surface. I floated to the center and exhaled, trying not to obscure the visibility with my fins. As the air left my lungs, I started to descend gently. The increasing pressure on my ears told me that I was dropping in the murk, my dive light still barely visible in the sulfurous tannic haze. My ears creaked and squealed as the pressure equalized, and I propelled myself with a small rotation of my ankle to move laterally toward the far side of the pond. Suddenly I pierced through the gloom into clearer water. At this promising sign, Paul passed me, spooling out a thin white lifeline that would show us the way home. We passed through a small gateway adorned with ornate hanging stalactites and I heard Paul yelp through his scuba regulator—the thrill that explorers feel when they are the first to uncover the unknown. The cave was a pristine time capsule. It was obvious that nobody had been here. The nearly transparent water seemed frozen in time. No disturbance. No garbage. No line beyond the point I had first explored. It was all ours to discover.

Following Paul, I quickly moved past the point where I had left my exploratory line from my discovery dive. With two of us in the cave, things were a little different. We were taking more time to carefully tie off and secure our safety line. Spending more time in place caused our bubbles to dislodge a lot of silt and worse; large mats of dirt and bacteria were now pummeling us from above.

Caves that have never been surveyed shed a lot of silt when the water is finally disturbed. Fins and bubbles both make a mess of things. It's almost as if the caves are sentient, mischievously opposing the first visitors and defiantly making it difficult to uncover their secrets. They lure you into the clear water, but you don't know you have trashed the place until you turn around.

With each breath, I was subjected to a steady rain of debris from the ceiling of the cave. Some chunks the size of a notebook tumbled from the ceiling, and as they broke apart, they obscured our visibility. I could see only the cave walls and floor for brief intervals in between the barrages of dirt and mud. Even Paul's light was just a dim glow in the murk, even though he was only one body length ahead of me. My quivering hand grasped the exploration line, and I wondered whether it had been a good idea to put two of us together in this new cave. For the first time, I truly felt nervous, and maybe it was because I was now beyond my training and experience. I took some deep breaths and tried to tell myself that Paul's expertise would protect us both.

As we continued to descend, the water became clearer. I relaxed a bit, seeing what looked like white stone castles through the murk, while Paul proceeded to unwind his reel into the unknown. We reached a depth of 130 feet and tied off our line to a spindly rock formation on the bottom. If the cave continued beyond here, it would take at least another day to explore. It was time to go back.

We carefully turned to retreat, and we each slipped a hand in a loose OK signal around the thin white guide line. We had to stay as neutrally buoyant as possible to avoid pulling and stretching our lifeline. If the line pulls loose, it can disappear into spaces that are impossibly small. We call them line traps, and line traps can become death traps if you can't find the way out.

The cave was a dangerous mistress. She was already luring me back for another look, deeper into the recesses of the earth.

Paul captured my attention too. Our carefree hike in the woods had been my idea of a perfect day. To me, there is nothing as good as being soaked with the sweat and mud of a great adventure. After that, I dreamt about exploring and about seeing Paul again. Before long I was spending every paycheck on camera gear and two-hour flights to Tampa, where I could train properly as a cave diver at Paul's shop in nearby Hudson. Although we were still flirting with a courtship, we were building a solid friendship, and I was excited each time he picked me up to head north for a dive in cave country.

CAVE COUNTRY

1993

CAVE COUNTRY IS not what most people envision when they think of Florida, but the region northwest of Gainesville is unique on the planet. Sitting atop one of the most prolific groundwater resources on earth, it is the antithesis of what I would have expected to find in the Sunshine State. There are no strip malls, few traffic lights, and more cows than people. It is the buckle of the Bible Belt, where you might find rattlesnake-handling preachers telling their female congregants to obey their husbands and fear the wrath of God. But if you are looking for the Garden of Eden, you just might find it in the greenway between High Springs and Marianna, Florida. Hundreds of springs boil up from the source deep below ground, spilling their crystal clear water into head pools that fill unspoiled rivers like the Santa Fe and the world-famous Suwannee. What makes Florida's springs so unusual among the aquatic wonders of the world is not what you see, it's what you don't. Most of the region's water flows below the ground through a dark and uncharted geologic wilderness. Lost rivers rise unexpectedly in one place, then disappear back underground in another; sinkholes and spring-fed lakes can be full one year and then, curiously, go dry the next. These hidden rivers are an explorer's dream.

The earliest human visitors knew this part of Florida was unique. Almost thirteen thousand years ago, Paleo-Indians, perhaps following a giant mastodon thundering through the sand-hill pines, were drawn to these bountiful water holes. In those days, glaciers to the north were just beginning to melt, and the Florida peninsula was much larger and drier than it is today. Surface water was scarce, and early peoples survived by camping and later settling beside the nourishing waters that welled up from the ground. Like an oasis in the desert, these springs were sanctuaries, attracting humans and animals to share their sustaining bounty. They continue to be the reason that North Florida became the hot spot for cave divers, who converged from across the globe to learn their craft.

Every drop of fresh water flowing through Florida's springs and caves originates deep underground in a vast reservoir known as the Floridan aquifer—the water source for more than 60 percent of the state's residents. Beneath a hundred thousand square miles of southern Alabama, Georgia, South Carolina, and Florida, this aquifer ejects more than nine billion gallons of water each day. Below ground, the water flows through voids in the sponge-like limestone as well as in between tiny grains of sand. Drifting down gentle slopes in the landscape, the water flows slowly. Where steeper drops occur, the underground rivers can rush forward with steady erosive force, creating enormous water-filled voids and tunnels.

By then a very experienced full-time scuba instructor and young explorer, I hardly reckoned that I needed to start my cave-diving classes at square one. Yet Paul insisted that I begin with an entry-level team by taking a cavern diving class. Cavern diving differs from cave diving in that the diver always stays close to the entrance, in the daylight zone of the cave. There is limited penetration, and the exit is visible just yards away. I had been in many caverns and thought there would be little new to learn. But even before the first day ended, I realized that I was illiterate in the skills needed to safely dive in an overhead environment.

One of the first instructional sessions included running guide lines in the grassy picnic area beside a watery blue gemstone in the woods called Ginnie Springs. With a tantalizing turquoise river only feet away, it was tough to spend an entire morning doing drills on dry land, dripping with sweat in the Florida sun. The water was calling. We took turns spooling out braided nylon line from a plastic reel, securing it around trees, barbecues, fence posts, and picnic tables, creating a webbed maze of lines to simulate cave passages. Walking through fire ant mounds and getting sharply burred grass seeds stuck to my ankles left me wondering if Florida was such a great place after all. Once I'd gotten the hang of using a diving reel with my eyes open, I was told to close my eyes and attempt to follow the line back to the beginning of the course without losing contact with the line and my teammates. Like a children's game of "blind leading the blind," we learned how to communicate through touch contact and effectively stay together as a group. As we slowly worked around the circuit, tripping and stalling when someone negotiated a hazard, it became apparent that teamwork was an essential proficiency. The thin line became our pathway to safety, and nothing was more critical than staying oriented to and connected to the exit of a cave. I recalled the poor visibility that Paul and I encountered in Cayman and knew that it would have been easy to get lost if I'd let go of the guide line in the murk.

My first effort to properly run a line inside a Florida cave took place at a doorway called Devil's Eye. I donned my toasty warm custom-made Canadian Brooks dry suit. The seven-millimeter-thick pink and black neoprene rubber exposure suit was overkill for the tepid water, but it was the only thing I owned beyond a tropical wetsuit that was suitable for swimming around warm-water reefs.

Dressing for cave diving is a balancing act. First you put on thick, insulating undergarments, and over those a dry suit to keep you warm and dry. The outfit can be stiflingly hot on land, but it's necessary to keep you warm on a long dive. But the suit makes you float, so you wear heavy lead weights that feel like they will drag you to the depths. A buoyancy "wing" that acts like a flotation device, and dry suit are inflated incrementally during descent to compensate for the depth induced compression that shrinks the wing and dry suit's volume. Without this virtual life preserver, a diver would cascade downward from the increasing negative buoyancy. It's a trick to finding the equilibrium between the weight of your gear and the lift of your wing. The result is a heavy costume that makes you look like an astronaut and walk like a knuckle-dragging Neanderthal.

Wedging a nine-pound lead weight in between my back-mounted twin tanks would help compensate for the buoyancy of my dry suit. Soaked with sweat, and anxious to cool off, I quickly dressed into my nylon-webbed harness and inflatable wing. I fastened a hefty twenty-pound light below my tanks, and backup lights to stainless steel D-rings on the stiff shoulder straps, bringing the total weight of my kit to well over 150 pounds. My knees felt like they were collapsing as I ambled to the wooden deck and the steps that led down to the spring run. Two young swimmers perched in the access route, toes curled on the lower step like small birds paired together on a branch. They squealed when their feet contacted the chilly water, flapping their arms and recoiling from the glorious wetness that would soon relieve me of being both overburdened with weight and overheated in the scorching sun. I asked them to move aside and I stumbled down into the natural pool. The relief was immediate.

My two teammates joined me in the translucent water, and we grouped together to review our safety drills. Cave divers prepare for a dive with an in-water safety check called an S-Drill. We examined each other's equipment from head to toe, listing off each item and ensuring its proper function.

"Mask?"

"Check. My skirt is tucked in."

"Regs?"

"I've breathed both."

"Me too. Thirty-two percent. Maximum operating depth one-twenty."

We rattled on through the list.

"Inflator?"

"Check."

Hiss. My buoyancy wing expanded as I pressed the plastic button and confirmed that it worked properly. "Wing dump works."

"Dry suit feed?"

We all reached to test the inflator mechanism mounted on the chest of our dry suits and set the over-pressure release valves on our biceps.

"Cutter?"

"I have one on my wrist and one on my shoulder harness," I said, touching each to make sure they were secure.

My teammates nodded. A young swimmer haloed in a bright orange inflatable tube drifted close and listened intently.

"Now, reels. I have the primary, safety, and a backup spool."

We all checked that our line reels were safely attached and the locking mechanisms were fastened tight.

"Cookies and arrows?"

The young swimmer began to laugh. "Cookies?" she said. "You take food with you underwater?"

I smiled and held up my plastic navigational markers—the cookies and arrows.

"How deep do you go?" the girl asked.

I said we were planning on hitting about a hundred feet of depth but had the potential to swim linear miles inside the earth.

She responded, "Is it dark in there?" shifting my attention back to our checklist.

"Backup lights, everyone?"

We all checked our two backup lights worked and stowed them in our harnesses for easy deployment.

"Primary lights?"

In turn, we each fired up our main dive light head, which attached with a long cord to the massive lead-acid battery beneath our tanks.

"Okay, let's do a bubble and manifold check."

We took turns ducking down just enough to reach back and check that we could access our tank valves, and we examined each other for leaks from the system.

"Turn pressure?"

We inspected our submersible pressure gauges and wrote down our limits on white PVC wrist slates. We needed to keep track of turnaround pressure so that we would leave enough gas for the exit, and emergencies.

"Tables and timers? Let's make sure we all agree on the dive plan."

My clunky gray Beuchat wrist computer came to life with a beep. Diving computers were not common at this time and rarely agreed with one another. We all carried diving tables so we could manually calculate the profile of our dive—the maximum depths and time—and manage any decompression stops we might need to finish before surfacing.

Finally, we were ready to descend and complete the safety drill with an air-sharing exercise. I dropped below the surface, feeling the familiar coolness seep into my neoprene hood. The water was as clear as glass, perhaps even clearer than the hot air above the surface. I swam to one of my buddies and slashed my hand across my throat, signaling I was out of air. He handed me the regulator from his mouth and uncoiled the entire length of the seven-foot air supply hose. We got into position side by side and rehearsed the drill while he retrieved his backup regulator. We then reversed roles, with me passing off my long hose.

Our fin tips merely grazed the bottom but stirred up the light clay and sand. The pristine beauty of the spring was temporarily reduced to a blizzard of silt. It struck me how quickly this unblemished water could explode into a dirty eruption of sediment. I knew I would hear about it in the debrief session with Paul after our dive, and vowed to be more careful with my fins.

With everything properly checked, we coasted down the shallow run toward Devil's Eye Spring. Quarter-sized freshwater flounder darted from beneath us as we swam over their hiding spots in the sand. Toy-sized musk turtles scurried as fast as their fat webbed feet would carry them to the submerged eelgrass and hydrilla plants obscuring the sloping banks of the river run. The towering cypress trees reflecting on the surface of the water gave me the illusion of swimming through a tubular hall of mirrors. I meandered beneath a bright pink figure-eight-shaped inflatable tube and exhausted the bubbles from my regulator. Muffled shrieks chattered above as I watched the teenagers' dangling feet propel the raft away.

We reached the edge of the chimney-like entrance to Devil's Eye Spring and peered over the brink. A small pile of sticks and tiny white shells were boiling around in the bottom of the hole, propelled by the flow of water from within the earth. At the time, I had no idea what the strong current caused by eighty million gallons a day of outflow would feel like, but I was tingling with the excitement of my first cave dive in Florida. I dropped down to twenty feet and fastened my diving reel to the worn branch of a tree that had fallen into the cavern zone. The branch was rubbed smooth by the hands of so many divers before me. I loosened the locking nut on my reel and made my primary tie-off before dropping into the first room of the cavern. Shining my light toward the source of the flow, I was met with a startling image of the Grim Reaper on a large sign that discouraged untrained divers from proceeding any farther. "More than 300 divers have died in caves just like this one," it cautioned.

I took a deep breath and clearly grasped how important this class was to my survival. I had already cheated the odds by exploring caves without training. Now I needed to become proficient. I did not want to join the pile of bones pictured at the Grim Reaper's feet.

As Paul watched closely, I made my secondary security tie-off with the line and pulled my body forward against the flow into the cavern's next room. The tough rock walls and floors closed in. There was an obvious path forward over the rocks, worn down by the scraping of divers over the decades. My twin 104-cubic-foot tanks screeched across the ceiling, and my chest skimmed along the floor as I fell down a claustrophobic channel to a narrow sandy corridor. Momentarily pinned by my tanks hitting the ceiling, I had to pitch my body back and forth and then tuck in my chin to free myself. I scraped my fingertips along the rock, skimming off a layer of flesh in the process. My heart raced, and I wondered if all cave dives would be this tight. We hadn't even gone beyond the cavern zone, and this cave had another thirty thousand feet of tunnels to offer ahead.

Trailing the reel line beside me, I knew I needed to find another spot to tie off before making a right-hand turn into a cobble-floor hallway. As I reached the turn, I was slapped in the face with a more powerful blast of current. I grabbed at a chunk of limestone, but it immediately lifted off the bottom while my legs pivoted around as if I were a weather vane in the wind. I raked my fingers across the stony bottom, never quite grabbing anything that was solidly attached to the bedrock. I tossed loose boulders aside and jammed my reel hand and elbow down into the substrate. My index finger was bleeding now, and I could barely inch forward as I kicked and pulled to move ahead. I ripped the skin on the rest of my fingertips; before this week was through, I would wear the cave trainee's badge of courage—duct tape covering each bleeding finger pad. I felt like I was swimming against the force of a tropical storm, and my false sense of bravado washed away in the flow.

My lungs heaved with the effort, and I was overheated, even though the water was a comfortable seventy-two degrees. I was soaked with perspiration inside my dry suit and restricted by its bulk from effectively moving along. Already exhausted, I had not even reached the point where I needed to tie into the cave's permanent guide line. There was a lot to this cave-diving thing, and I was beginning to doubt whether I was going to be able to do it.

The cave diver's watchword is that "anyone can call a dive for any reason," meaning that whenever you want to leave the cave, all you need to do is turn to your partners and offer a thumbs-up sign. There are no questions asked. The other divers respond with the same hand signal, and everyone swivels around and retreats together.

On this dive, I had reached a location in the cave called the Lips, where an enormous vertical gallery makes a right-hand turn into a gushing torrent of water coming out of a low horizontal slot. My partner behind me flashed his light beam back and forth to gain my attention. Even though I was enjoying the beauty of this underground wonder, I was hot and tired, and I couldn't have been happier to see him thumb the dive. Relieved, I turned around and began to race toward the entrance in the slipstream of rushing water. I watched Paul beside me, yielding to the flow and drifting without expending any energy. His ankles flexed and sculled, rotating his fins to make a turn. He looked over with sparkling eyes, revealing the joy he feels drifting through the conduits inside the earth. My doubts about cave diving vanished. I was intrigued—spellbound by the challenge and glory of the cave and fascinated by this graceful being, silently drifting through these caverns beside me.

After more than a year of jetting between Cayman and Florida, with Paul's steady mentoring and training, I earned my certification as a Full Cave Diver. We still hadn't found a romantic connection, but we had become good friends and trusted diving companions. Although I was almost comically taller than him, I was attracted to Paul's strong masculinity and engaging smile. He struck me as being the kind of man who could be gentle but also defend himself if needed. I found that attractive. He was also the catalyst that was helping to refine my interest in exploration. Every time I dived with him, I learned from his decades of experience. Still, although we spoke on the phone much more frequently, our conversations were always centered on diving.

Meanwhile, my underwater photography efforts were starting to pay off. I managed the advertising for the Diving Lodge, sold copies of my photos of colorful walls and curious turtles to *Skin Diver* and Rodale's *Scuba Diving* magazines, and had my first articles published in the popular monthlies. I led the Cayman Islands delegation to the annual dive industry trade show, organizing twenty-eight display booths, charity auctions, and reef conservation events. I was making a name for myself in the business.

In the fall of 1994, I got a life-changing phone call. Still intent on matchmaking, Danny, who had since moved back to Florida, told Paul that I was looking for new work opportunities in diving, and Paul gave me a call. "Jill, would you consider watching over Scuba West for a couple of months?" He had been invited by Dr. Bill Stone to participate in an expedition in central Mexico and he needed someone he could trust to keep an eye on his scuba store. "You can live in my trailer behind the store and cover for me while I'm away."

Bill Stone was an icon in cave exploration and engineering, a brilliant man who was equally comfortable in the field and in a lab. I had read his book about a project that had pioneered technical exploration and mixed-gas diving—a type of diving that requires complicated electronically-controlled life-support gear called rebreathers and lengthy decompression stops using fluctuating mixes of helium, nitrogen, and oxygen. Bill built his own rebreathers, tested them in some of the most challenging caves on earth, and planned to take his gear to outer space. Caves were his proving ground and his expedition partners were his test pilots. I was in awe of his background and his dreams.

Knowing that would be a prestigious expedition for my friend, I wholeheartedly supported it. It was a win-win situation, giving me just the headspace and time I needed to figure out how I could profitably combine my creative design background with my growing diving abilities. I knew I couldn't progress much further in Cayman. If I moved to Florida, I could send out résumés and still take time to cave dive, shoot photos, and plan the next phase of my life. Although watching the shop wouldn't pay me any-thing, I also wouldn't have any expenses to worry about. It was easy to say, "Yes!"

Both in cave diving and in life, the darkness of uncertainty was beckoning. I was scared, but I knew that if I could be brave enough to step over the brink into the blackness, my eyes would adjust and new possibilities would be revealed. I had moved to a new country once before, and it felt easier the second time around.

THE DEEPEST

1995

I ARRIVED AT Tampa airport on an unseasonably hot afternoon in January, prepared to look after Paul's dive shop and perspiring under the burden of three big duffel bags of diving gear and a few clothes. All of my other belongings had been placed on a "curb alert" in East End, Grand Cayman. My friends in the village quickly scooped up the remains of my life on the island—a trusty bicycle, a thick comforter, and miscellaneous household goods. I was starting something new and exciting, and the vibration of anticipation was back, pulsing in my chest. The shifting transformation of my life was exhilarating but nerve-racking too.

I returned to Canada to pick up my old Toyota Tercel and a few warm clothes, and Paul invited himself along. The four days of car travel gave us the chance to kindle a little more interest in each other, but I felt myself holding back from romance. He was older, but more crucially, he lived in the U.S., where I only had visitor status. We shared hotel rooms and even beds, but there was still some indiscernible barrier between us. I liked him a lot, but I was afraid to start a relationship that could be torn apart over nationality. I was also concerned that he had a son. I was not sure if I was ready to contribute to the upbringing of a young pre-teen boy. Although Paul seemed to have a friendship with his ex-wife and her husband, I was afraid I would just complicate things.

On the way back south, we stopped in Gaithersburg, Maryland, to meet Paul's expedition buddies, including Bill Stone, and help prepare some of the gear that the group would take to Mexico. Since I had nothing on my calendar and no place to call home, I was glad to tag along as long as Paul and the team needed help.

We arrived at Bill's home in an unpretentious suburb near Washington, D.C. It was clear that Bill had other priorities than maintaining a homestead. The lawn needed mowing and the decor was spartan, but this towering man with an angular face made me feel right at home, and we immediately hit it off. I also connected at once with his live-in caving partner, Barbara am Ende, a strong and witty PhD. It was clear to me that she was an equal teammate in planning this expedition. I admired the way she managed to balance a great career as a geologist with her passion for exploration. I hadn't met many women in cave diving and I immediately wanted to know her better.

Touring Bill's garage, an inner sanctum crowded with diving equipment, I smelled something that reminded me of campfires and drying mud. I saw that some of his scuba regulators were still caked with thick brown clay from his last expedition. I scooped up the tangle of hoses and mouthpieces. "Do you want me to service these regs?" I asked. "I'm happy to clean them up." He looked pleased.

By evening, the four of us were cooking up a delicious pasta dinner. We talked politics and global issues. I sat on a folding camp chair while Bill strummed popular rock melodies on a vintage electric guitar. The room was mostly empty, further evidence that Bill and Barb's priorities resided in buying climbing and diving gear rather than furniture. I begged for details about the project they were launching in a few weeks.

Bill's wrinkled face exposed the determination of a man who had been living on the cutting edge of exploration for more than twenty years. His temptress resided somewhere below the mountains in central Mexico, in what could be the world's deepest cave. He used the Sierra Mazateca mountains as a proving ground for radical life-support systems that might eventually head to outer space. Stone, am Ende, and other cave explorers spent all their vacation time and extra cash surveying a jigsaw puzzle of passages descending through the remote mountain platforms that carried cascading waterfalls the height of skyscrapers. Using miles of rope and sometimes dive gear, they traversed rivers inside the earth that carved canyons and roaring raceways through lightless vaults of rock.

I had read about their recent expedition in *Outside* magazine and asked if they felt comfortable sharing the details. The portrayal of Stone in the article wasn't pretty. He had been painted as a task-master who pushed his team too far. Just one year earlier, on an expedition sponsored by National Geographic, Bill and Barbara had descended from the mountaintop through the inner workings of the planetary plumbing to reach a point deeper inside the earth than any human had been before. This type of exploration, called sump diving—alternating between dry cave passages and completely submerged tunnels—requires strong leadership and faultless equipment. Their small team had worked for three months in the darkness, exposed on steep, rocky pitches, staging tons of high-tech equipment while camping in the musty chambers where you can't even see your own hand without turning on a flashlight. They bivouacked for weeks at a time in the cave before heading out on dangerous dive excursions in hope of finding more dry passages and a way downward. The dives pressed team members not just to their absolute physical limits but also to the psychological edge. And that was just to get into the cave and conduct new exploration. The greatest challenge of this type of deep caving includes saving enough gear, resources, and personal grit to get yourself out.

At the deepest point of penetration, beyond more than two miles of rope, their teammate Ian Rolland died during a diabetic emergency. Surfacing alone from a dive through a flooded passage—a sump—he had reached the end of his energy reserves and fell unconscious in shallow water. He drowned when his rebreather mouthpiece fell from his slack jaw. Bill, Barbara, and the rest of the team toiled inside the hellish mountain darkness for twelve days to recover the body of their friend. Rolland left behind a loving wife and infant son. (In adulthood, his son went on to make record-breaking dives in that same cave, following in the footsteps of the father he would never know.) But for Bill and Barb, the news only got worse. As they recovered Ian's body, they learned of the death of their close friend Sheck Exley. The absolute pillar of the cave-diving community and recognized authority on the rules of safe cave diving, Exley had died attempting to break a deep-diving record in northern Mexico.

After holding a simple memorial service for Ian, Bill and the team forged onward, plummeting even deeper into the system to finish their survey work. Some journalists criticized them for continuing the project after the fatal accident, but team members who stayed on did so voluntarily. They felt they had a responsibility to Ian to finish their work and de-rig miles of rope and gear from inside the mountain before heading home. Leaving behind ropes and equipment would not just constitute bad form but would also create a hazard that might lure an untrained caver into the darkness where risks were beyond their comprehension.

But even with grief fresh in their minds, Bill and Barbara were ready to return to Huautla, Mexico. This time, they wanted to find a record-breaking route through the cave and declare it the deepest in the world—the greatest vertical distance from entrance to exit. But this time, rather than start at the top of the mountain, they wanted to work from the bottom up. I recognized how much their "hobby" was a high-stakes game, far more than a simple pastime. Listening to their stories, I felt a thrill, a mix of infatuation with Paul and a desire to be part of something big. I wanted to work on a team that was worthy of coverage in *National Geographic*.

Over breakfast the next morning, Bill asked if I would like to join the project. I wanted to blurt out, "Yes, yes, yes!" Yet as excited as I was to join the expedition, I was uncertain. Could I truly add value to their team? Plus I had agreed to babysit Paul's dive shop; I could hardly drop that commitment and leave him without help. As I opened my mouth to reluctantly decline, Paul interrupted and said, "I think you should come with us."

In one sentence, my life was transformed. Seeing in me a worthy diving partner, Paul opened up a world of exploration. This would be my first real expedition. It was an incredible opportunity. But his vote of confidence also showed me that he truly cared about me. I could have danced on the ceiling!

JUST GETTING TO Huautla was not for the faint of heart. We loaded up Paul's fifteen-year-old Volkswagen Vanagon from floor to ceiling with dive gear and left Hudson, Florida, on a warm evening in the spring of 1995. We took turns behind the wheel, stopping only for gas, food, and roadside naps until, twenty-nine hours later, we laid our heads on the last crisp, clean sheets we would know for another month. As the sun warmed the day to a muggy heaviness, we showered and tidied ourselves for our appearance before the border guards. The tired VW camper's odometer rolled over 160,000 miles as it knocked and struggled its way to the checkpoint.

Just the other side of the border, we stopped to pick up a local man.

"Do you speak English?" I asked.

"Sí," he responded.

I tucked a ten-dollar bill into his hand and requested, "Guide us out of Matamoros?"

The border town had a well-earned reputation for being dangerous. Paul told me it would be much easier to transport a Spanish-speaking guide who could help us through the numerous military checkpoints. I felt nervous, but our new guide had an easy smile that comforted me. He climbed up into the passenger seat as I slid over toward Paul. I had to perch on a small box of gear we had wedged between the seats, my head striking the ceiling with every bump.

From Matamoros onward, it was best to follow a bus or a truck and let them blaze the trail. They knew the roads far better than we did and would slow for *topes*, or speed bumps. The experienced drivers knew when the road ahead was broken or washed out.

We drove for fourteen hours, snacking on food I had stashed in the van's small fridge, until Paul started nodding off at the wheel. He had a terrible habit of insisting on driving well beyond exhaustion. I talked to him, sang, and blared Mexican folk music on the crackling radio to keep him awake. I slugged him on the bicep, but he would not surrender the wheel; he didn't like being a passenger, arguing that it made him feel carsick. I was feeling queasy, too, but with fear, as I watched his eyes flutter and his head bob. Finally, I persuaded him to pull over and camp for the night. We found a quiet pull-off where a couple of farm vehicles were parked. The night was thick with a tropical steaminess. Lush green growth mixed with a subtle aroma of the oil business, which was prolific in the region. Our little van had a pop-up loft, and we crawled into bed to get a few hours of sleep before the hot sun disturbed our nap.

The height of excitement came the following day as we ascended from Ciudad Mendoza up over the mountains and down to Tehuacán. The road rose into the clouds with over half a mile of steep, tight hairpin turns. With the speed of a lazy snail, it took us two hours to conquer this single, stubborn climb. From there the road deteriorated further. A single rutted track without guardrails offered a startling view of trucks that had failed to negotiate the tight turns. Burned-out hulks were strewn down the mountainside in tangled clusters of steel and rubber. At times we met oncoming traffic and cut so close to the edge that I thought we were going to drive off. On the passenger side, I could look down the perilous drops and see our tires precariously close to the lip of the road.

Our final approach into the mountain town of Huautla took another four hours. At times the turns were too tight for the arthritic van, and its antique joints were pounding, metal upon metal, at every curve. Sometimes we were forced to make three-point turns. We had to back up, regroup, and inch forward through the switchbacks. At one point, the brakes were heating to the point of failure, smelling so bad that it turned my stomach. I held my breath, looking for a safe place to pull off to let the brakes cool.

Close to our destination, we split a tire open and had to scramble to repair it, worried that an oncoming car would not see us and hit us. *Curvas Peligrosas*, read the road sign over my head. Dangerous turn, indeed. I crouched, my full strength coaxing the tire iron against seized lug nuts. My nerves were as taut as piano wire.

Four days after leaving Florida, we finally reached the village of Huautla de Jiménez. Set into the mountainside, the town was laid out along switchbacks sprawling through the lush high jungle. It was a sultry, humid paradise, abundant with fruit trees and cackling, colorful birds. As the sun set, the town was coming alive in the foggy evening dampness. Children at small thatched roadside stands held up bags of peeled oranges, hoping for a quick sale. Kids were playing basketball on a court beside a two-story-high mural, a magnificent painting of Jesus on the cross. Purple mushrooms were growing out of his upturned hands; at his feet, a carpet of wildflowers and more of the region's psychedelic mushrooms.

Walking through the alleys thick with charcoal smoke, I smelled a spicy blend of copal incense, tortillas, and barbecued meat. A small television hung in a doorway facing the alley. Two men sat in broken white plastic chairs while a gaggle of children climbed on top of each other for a glimpse of Jean-Claude Van Damme on the screen. Women in crisp white embroidered indigenous dresses beckoned me to their market stands to buy vanilla beans and honey. Was it my height or my pale complexion that was making them laugh?

Finally, my eyes settled on my friends. Towering above the Mazatec locals, Barbara am Ende and Bill Stone drew a crowd of fascinated onlookers. We hadn't even made it through our welcoming hugs when Dr. Noel Sloan, our superstitious team physician and fellow cave diver, launched into a story about their getting shot at in the night. The three of them had had to evacuate base camp when gunshots rang out over their tents. They scrambled into the jungle in the dark. Some hours later, they crept back in the moonlight to discover only a single climbing rope was missing. A good rope was as valuable as gold in these parts.

As expedition leader, Bill had spent the day speaking about the gunfire with the mayor and other local officials. Without their permission and blessing, our exploration could not take place. We needed to ensure we were welcome before returning to the camp, which was a day's hike down the mountain.

The next bit of bad news arrived in the form of a landslide that occurred as the rest of the team was coming up from the Santo Domingo Canyon to meet Paul and me. Early seasonal rain washed torrents of mud into the gorge, destroying the clarity of the water at the cave resurgence—the point where the water that runs through the inside of the mountain re-emerges—and making diving next to impossible. Plans had to change, and we needed to improvise. We had two vehicles full of diving gear, compressors, and lights and might never get to use any of it. All that careful packing and preparation and now the equipment might remain untouched. My first real cave-diving expedition could be ending before it started.

We decided to trim down the gear to prepare for a two-week attempt at fully exploring the canyon and any caves that could be found there. With luck the river water would clear and we could dive, but in any case, we would probe every crack and crevasse we could find in the rock faces. Our goal was to find a direct route from the canyon floor to the deepest previous exploration inside the mountain, and hopefully a dry bypass that might be a shortcut for cavers who otherwise had to come down from the mountain-top, a route that required a treacherous vertical mile of rappelling, crawling, and diving to arrive at a place deeper inside the earth than anyone had ever ventured. But setting up ropes and base camps inside the mountain took weeks, and there was always the risk of floods and rushing waterfalls trapping a team. That's why working from the bottom up was appealing.

Our plan was to park our vehicles at the small village of Río Tuerto and descend on foot down nearly six thousand feet to establish a new base camp near an underground river that spilled out of a cave. Harnessing fourteen burros and enlisting three strong local men, we could transport all the equipment in a single day. We would take enough scuba gear for Paul to make an initial scouting dive and one set for emergency rescue—more likely, recovering his body. Just stuffing the extra gear into a pack sent chills down my spine and reminded me of the dangerous nature of expeditionary work; there are few rescues in cave diving.

My own equipment would stay on the mountaintop, because I was not ready to conduct a rescue or recovery. My loosely defined job was to support the expedition in any way necessary. I would guard camp, purify drinking water, cook, and gather firewood. If all that went well, I might get some training in single-rope climb-ing technique and be able to help survey dry cave passages. I was excited, and up for anything. I was on an actual scientific expedi-tion, led by some of the world's top experts.

The next morning, after an evening of downing small glasses of sugarcane-derived moonshine called aguardiente, we picked up a colossal cache of equipment that had been stored for a year at a tortilla factory and edged down toward the village of Río Tuerto. The grueling transit was only a few miles, but we pushed the limits of our vehicles' brakes and burned up the rotors. Washed-out pot-holes and large boulders threatened to crack the frames of our vehicles as we bounced and lurched over the massive craters. And then, within two hundred yards of the day's destination, the road was blocked by a truck stripped of its tires and raised up on cinderblocks. There was no way to get around it, and so we unpacked our gear and continued on foot. On the other side of the obstruction, a large crowd of people came toward us, inquisitive about all the strange equipment and our unfamiliar faces.

As fate would have it, the gathering was a wedding celebration. Barbara and I were swept into the crowd of revelers making their way along the dusty clay road. A playful-looking elder with a grin pulled at my hand while commanding in the Mazatec language that I join him to dance. I wasn't sure whether to join the party or bow out and apologize for interrupting their festivities. But as I hesitated, it became clear that this celebration had a force of its own. The men on our team were given beers, and Barbara and I danced with the elders.

After a few dances, lots of laughter, and lots of handshakes, we collected our equipment and carried on. Some of our new friends from the wedding helped us hoist our gear into the loft of a nearby barn. This would be our last rest stop and final chance to minimize the equipment that would accompany us to the valley below.

The wooden floor of the barn was dusted with a light sheen of pulverized livestock dung. Giant rats scurried around us, unfazed by the bustle of human activity. I tried to act brave, but was unsettled. We agreed to pitch our tents in the loft to isolate us from the critters and their potential contagions.

Five days into my trip and not yet having had a single full night of sleep, this tent was the most comfortable bed I had ever climbed into. I was exhausted. The incredible remoteness of where we were felt surreal. Here I was, experiencing people and places I could never have imagined. There was simply no connection with the world beyond a one-day walk. Televisions and telephones were a day's journey away. Goods and services were bartered, and few people had a need for money or technology. This happy community lived in isolation and solitude, untouched by outsiders and, as far as I could tell, unaware of the international news cycle.

I had just drifted into the blur of a dream when a shrill sound startled all of us awake. It turned out that Charlie, the chief burro, was now sharing the barn with us and bellowing like a band of amateur bagpipers, just one thin floor below our tents. His howling hee-haws kept up all night, until just before daylight, when a cacophony of roosters began to compete. So much for starting the day well rested.

AT DAWN, A band of muscular porters and more bellowing burros positioned themselves by the barn, ready to take some of our gear down the mountain. Double tank sets were balanced on either side of the beasts' saddles. A small portable compressor was lashed to another. Colorful plastic baskets held a few precious fresh vegetables nestled alongside water bottles containing pulverized freeze-dried food. Chicken stew, vegetable beef soup, and mashed potatoes were desiccated and compressed to minimize size and maximize efficiency. We trimmed down our personal packs for the day-long hike. It would be hot, and the trail would be arduous. Heat exhaustion and dehydration could be significant problems in this remote terrain.

I was a veteran hiker, but when I tried to shoulder my pack, I couldn't even lift it onto my thigh, it was so heavy. As the newest member of the team, though, it was important to me that I look tough and capable, so I sat on the ground, slipped into the shoulder straps, and strained from a squat to an upright position. Afraid I would not be capable of getting up a second time, I vowed to leave the pack on all day. If I had to rest, I would find a place to sit upright and bolster the pack. As the day wore on, though, my pack seemed to grow heavier, and by the time we reached the canyon floor, my toenails had paid the price of jamming up against the ends of my too-snug boots. They were bruised and falling out, and within the next few days, most of them would be gone. It was excruciatingly painful, but I couldn't appear weak to my expedition mates, so I masked the pain, determined to carry on, unflappable. I wanted my new friends to know they could count on me. If I became a burden to the group, unable to pull my weight, this would be the last time I would be invited on a project. Perhaps I felt additional self-imposed pressure as a woman. In the past I had been held back from opportunities because I was a girl. Even though nobody in this group had expressed sexism, I was determined to be as strong as any of the men.

The path was narrow, and the scenery was breathtaking. As we made the climb down, the environment morphed from a moist rain forest to blistering-dry dustiness. The rocks on the cobbled trail rolled under my feet, causing me to slip and brace myself on tree branches along the steep path. We hiked for hours until, finally, four hours in, with a face as red as a beet, Paul wearily reported that he had no energy left. We had drunk all our water, so when we reached a small stream, Paul crumpled to his knees, pursed his lips, and began to sip between the tumbled rocks. I told him that he shouldn't drink that water. With the burros and other animals guzzling and shitting in the stream, it was sure to cause him intestinal grief. But there was no stopping him. He lay in the water and lapped it over his head to cool down. I stood leaning against a tree, afraid I wouldn't be able to get up again if I sat down. I was still trying to impress Paul and the other experienced expeditioners. "You are capable and strong," I tried to convince my doubting body.

Once he'd revived himself, Paul was back on his feet, green algae now dripping from his curly hair, and we set out to hike the last mile. We were finally on the canyon floor, at least an hour behind our more capable teammates. We eased along the bank of the Santo Domingo River, perpendicular rock walls looming over the shaded water that ran swiftly over shallow rapids. The occasional fall of a rock from the dominating cliffs punctuated the noise of rushing water. We used the smoother rocks and boulders to hopscotch up the river, and it was impossible to keep our feet dry. I knew that long swims in wetsuit boots could lead to ulcers and painful sores on the feet, but I hadn't even had the chance to dive on this trip and the skin was already sloughing off my feet. The needles of chilled water stung through my boots, and I could feel my blisters tearing open.

We finally arrived at a sandy beach where Bill, Barbara, and three other teammates had already pitched tents for our base camp. It was the first flat surface I had seen on the riverbanks. Our friends were relaxing around a roaring driftwood fire that was set to discourage the relentless mosquitoes and flies from stinging. Paul and I picked a spot for our tent, and after I'd rolled out my sleeping mat, I inspected my shriveling pale, pink feet with my headlamp and bound them up with moleskin and antibiotic cream. Even though the sun had set, the evening still felt like a sauna. I slept like a baby, but Paul dozed outside, his legs half-submerged in the muddy river. It was simply too hot for him.

The next day, my long-lost sense of humor was restored and my enthusiasm for new adventure eclipsed the aching discomfort of bad feet and sore muscles. We scrambled four hundred feet upstream in the canyon to look at the portal that served as the lowest exit of the Sistema Huautla cave system. We waded upstream through the fast-moving river, listening to the buzzing of cicadas echo off the canyon walls. Their shrill chorus mixed with a loud wind that brought with it clouds of clay dust from the mountaintop. I was enjoying our surroundings when, suddenly, out of the dust emerged a skinny burro and four men on the opposite bank of the river. They were armed. I immediately thought about the gunfire that Barb, Bill, and Noel had experienced the night before we arrived, but I soon recognized that these scrawny hunters were looking for food, not trouble. The Mazatecs peered at us from across the water, unable or unwilling to respond to the questions we shouted at them in our best attempts at Spanish. We waved, smiled, and when they acknowledged our peaceful greeting, we carried on.

We reached the opening to the cave, a roost about fifteen feet above the river, gaping up toward the sky. A single barren tree poked out of the flowing cave spring, growing in the water that filtered down through the interior of the mountain. Bill had described the cave resurgence to me in Maryland, but I wasn't prepared for how unremarkable it actually looked. This single hole was the spot where the entire contents of a mountain full of waterfalls poured out. I couldn't imagine how this gateway the size of a large garage could be the back door to what might be the deepest cave in the world. I don't know what I expected for a dive site, but this was not it.

The rest of the team left Paul and me while they climbed to a spot called Narrows Cave to explore and survey until they were too tired to continue. These forays of exploration and mapping, taking up to thirty hours of effort at a time, would yield fresh geological and scientific knowledge of places never before documented. Meanwhile, I would help Paul shuttle gear down into the cave and wait for him while he went diving. I would be alone to deal with any emergencies that might arise; what would I do if our friends didn't return from Narrows Cave?

If you cave dive long enough, you will eventually face the death of a friend. Worse, you may even recover the body of one, or hold them as their life force ebbs. In those moments, your life will be changed forever. Back then, in Huautla, I was new enough to cave diving and exploration that I had not yet lost a close friend. In my gut, I knew that if I were going to participate in extreme endeavors like this expedition, my days of innocence were numbered.

While Paul was busy assembling regulators on tanks and preparing his dive slate—a waterproof note tablet—for survey,

I made several trips to a spring to collect drinking water for the camp, purifying it through hours of careful manual pumping through a fine filter. Although my arms and wrists were tiring, it was worth it to have clean water available and avoid the gastrointestinal bugs that would have slowed the team. Each time I made a hauling trip from base camp, I had to ford the river twice, clambering over wet rocks to reach the opening to the cave four hundred feet upstream. My skin was dry and covered in silty dust and my throat felt like sandpaper. By the time I completed my last trip, Paul was finally ready to dive.

I crouched by the water's edge inside the cave and wished him good luck. His masked face and more than two hundred pounds of perfectly balanced scuba gear vanished into the murky water. Then I was alone. At first, all seemed well and peaceful, but while I sat enjoying the coolness of the cave, I observed the flowing water ebb and then halt. The arched ceiling above me disappeared into darkness, but around my feet I noted large tree branches that were wedged in the broken dolomite blocks. I could barely make out dirty scum lines up on the walls, left by much higher water levels. I couldn't imagine what would cause this stream to abruptly stop. It made me very nervous.

Minutes later, the river outside the cave came to life in a startling roar. I jumped from my perch and climbed out of the hole as streams of mud flowed down the canyon, filling the riverbed with water the color of milky coffee. The flow accelerated as the river level rose over a foot in just a few minutes. The tranquil stream reshaped into a rush of rapids, flowing fast over the rocks, bringing branches and debris racing along with it. This was nothing like I had ever seen, and I didn't know whether to be frightened or exhilarated by this thrilling display of the mountain's power. Nauseated with fear, I poked my head back into the cave, hoping Paul had returned. I yelled his name into the darkness and heard nothing back. I was learning a lot about this nondescript hole in the wall. The canyon and the cave could quickly transform from friend to foe. I felt completely helpless waiting for Paul to return, but as I sat by the water, the river slowed and calmed me down. I wasn't sure whether there had been an avalanche or mountain rain, but either way, I knew that Paul would be returning in very bad visibility. The water looked like cappuccino.

When he finally surfaced more than an hour later, I blurted out, "What happened?" One of the things that both attracts me to Paul and maddens me about him is his unflappable coolness in the face of adversity, and that moment was no different. "The vis was terrible," he said without fanfare, "but no big deal."

The following night, all the muddy adventurers returned triumphantly to base camp. After digging through a small hole that was plugged with clay, Noel and the rest of the team had found new passages. Noel had squeezed through a tiny cleft that tugged his pants down to his ankles, and when he backed out again, he declared that strong air had been hitting his face. That meant that more cave passages lay beyond. Paul had success, too, breaking out of the murk into clear water and surfacing in an air-filled chamber beyond the water-filled sump. I also had a small triumph for camp comfort when I came upon a large slab of wood in the river. Paul and I floated and surfed it back to camp and set it up as a rustic dining table.

That evening, while everyone else cleaned off after the day's adventures, I prepared reels of knotted line for surveying the cave passages, and I made one more trip to the drinking-water spring. With the stronger current and higher water levels, it was now an eighty-minute round trip, carrying back five gallons of water weighing over forty pounds. It was hard work, but it had to be done every day to support the team.

UNDER THE DAWN light the following morning, I sat on the sand on the edge of the river. On the far bank, a steep wall rose straight up out of the water. Bounteous bushes and vines grew out of the sheer face, creating a playground for a gymnastic creature that tumbled through the branches. The coatimundi had an otter's face, a raccoon's body, and a long, curled tail like a monkey's. I watched it drop acrobatically to the water to drink. Paul emerged from our tent, hunched over and groaning, walking like he needed to be oiled—the descent from Río Tuerto had caught up with him. My muscles, on the other hand, were wide-awake, but it was my feet that felt like they were ninety years old.

While eating a breakfast of oatmeal and raisins, we discussed Paul's dive the previous day. He had explored nine hundred feet of passage to reach a point where he could surface inside a dome filled with air. The beginning of the tunnel was narrow and cramped, but as it enlarged, the water became clearer. He was eager to go back to see if he could find a way onward into more unexplored territory. So, while most of the others headed off to explore more dry caves, I shuttled Paul's gear and full scuba tanks to the cave. Noel and his girlfriend, Kim, hunkered in a hammock to rest and keep an eye on base camp.

Paul's dive plan included a ninety-minute bottom time—the time spent at depth before beginning a slow or staged ascent—laying guide line and surveying on the way out. When he finally left, I occupied myself with "housekeeping" measures to make the cave safer. I moved wobbly rocks and built a clear path so it was easier to get out of the cave with our heavy double tanks. I swept off a small shelf of rock above the high-water line where I could stash Paul's dive gear, because it seemed pointless to lug fins and other items back to base camp after each dive. Then I stripped and took a bath in the cave. It felt great but was somewhat pointless. My clothes were permanently fermented in sweat and were taking on a mildewy funk. I would throw them away before I got back to Florida.

Ninety minutes passed quickly, but time slowed down when Paul was late. As the two-hour mark passed, I began to get concerned. He was now thirty minutes overdue, but I knew there was a small air pocket on the route where he could surface and breathe, and that kept my worry at bay. As the minutes rolled by, though, my heart rate increased. How long should I wait before I took action? If a diver was thirty minutes overdue from a dive in a Florida cave, I would call in an emergency, but this was different. There was nobody I could call who could do anything other than worry with me.

I didn't feel right leaving the cave, but decided I should make the twelve-minute hike back to base camp to alert Noel that Paul was missing. Noel would know what to do, given all his experience. As I stumbled along the cobbled riverbed, I noticed the creek was even muddier than before and that the water was gaining momentum again. I arrived at camp half-winded and found Noel and Kim still in their hammock, but naked. I quickly turned around to give them privacy, and told Noel that Paul hadn't yet returned from his dive. "I think we should carry more gear to the cave in case we need to search for him," I said.

To my surprise, he laughed. "He's one hour overdue? Jill, I once waited four *days* for Bill to reappear from a dive. Relax!"

I was mortified. Not only had I interrupted their romp in the hammock, but I had also proven myself a complete rookie. I waded back across the rock-strewn river crossings to the resurgence, hoping to find Paul materializing from the water. To my disappointment, the cave was still eerily vacant. Now he was two hours overdue.

It was at that moment that I noticed the world around me had gone silent. The cicadas were noiseless, the birds in the trees hushed. Even the moving water in the cave entrance was still. I crawled up to the mouth of the cave to see what was happening in the gorge. It was completely creepy, as if the earth had stood still.

Then I noticed a faint rumble, first feeling it in the pads of my feet. Gradually it intensified, crawling up my body in goose bumps. I heard a thunderous roar in the distance, and it was amplifying. Then I saw a front of muddy water crashing down the gorge toward me. Rain in the mountains had let loose a torrent of murky water that was making its way toward us in the form of a landslide, and the gentle creek outside the cave quickly morphed into a cascade of clay-colored water filled with branches and debris. I crawled down into the cavern to inspect the spot where I had said goodbye to Paul. The water inside the cave was rising too, inching up the bank where I now knelt. I had to do something.

I needed to report this new development to Noel. I crawled up out of the cave and jumped down into the river. It was no longer ankle deep—in a matter of minutes it had risen up to a height above my knees. My legs were tugged from beneath me and I was swept downstream. I grasped for something to hold on to and directed my swimming efforts diagonally down the river. Feeling the pebbled bottom grating my knees, I bounced off the rocks, using my already bruised feet to dig in and attempt to steer myself toward the shore. I was out of control, coughing and spitting water in between attempts to get a dry breath. Finally, I was able to wrestle my way to the shallows. I crawled ashore and scrambled down the bank, choking up a mouthful of vomit and dirty water. I had to cross the river one more time to reach camp, so I launched my body back into the fast water far upstream from camp, hoping to ease over to the opposite bank and beach myself near the tent site. As I got close, I watched as my fragile nylon tent bent against the rising water and broke free of its pegs, which were loosely held in the sand. My tent floated downstream just as I washed up on the sandy bank.

"Noel, we have a big problem!" I yelled over the din of the rushing water. He awoke from his postcoital nap, surprised to see the camp rearranged: ten feet of beachfront was now underwater and the fire pit had washed away. "Paul isn't back and something has happened upstream! I think there was a mudslide!"

Noel instantly transformed into a man of action. He hurriedly gathered his dive equipment, and without waiting for him, I headed back upstream to the resurgence with a backpack of gear. I recalled Bill's advice to leave the chest strap on the pack unfastened. Although fastening the strap made the fifty-pound set easier to carry, he had warned me that if I fell in the water, I could drown if I was unable to get free of the pack. That risk was all too clear after my inadvertent swim just twenty minutes earlier. Though the water had, thankfully, returned to shin-deep level, it was still a fast-moving muddy stream, and my adrenaline was blazing.

I vaulted into the cave, looking for signs of Paul. Bubbles on the surface would show me that he was on a safety stop. I dropped the first backpack full of dive gear and noted a faint beeping below me. I looked down to see a hand break the surface of the cloudy water. It was Paul's. For a moment I thought it was his body, being spit out by the flow of the cave water, and my heart slammed against my sternum. But then his arm moved with intention and slipped back beneath the surface. He had returned to the entrance with an obligation to make a very slow ascent to decompress his body but, because of the muddy conditions, he couldn't see the display on his computer, and because he wasn't able to surface immediately, he was using the audio alerts on his wrist-mounted dive computer to inform him about each stop he had to make on the way up. I didn't know how long he had to stay underwater, but at least I knew he was alive.

It was in that moment that Paul and I became forever tied. I had never been so worried, and the thought of losing him made me feel different about him. It would have been the greatest loss of my life. I realized how much I really cared about him. He had opened a new chapter in my life, and I was falling in love.

While Paul finished his decompression beneath the surface, I had to call off the alarm. The river was almost back to its normal depth and flow, so I was able to hop quickly along the uneven rocks. I met Noel halfway up the gorge. He was carrying a set of double tanks. "He's okay!" I yelled joyfully, the words echoing off the canyon walls. With a kind, knowing smile, he hugged me. He had seen the worry in my eyes, but he also had a lifetime of wisdom I had yet to gain. Whether because of his long background in caving or his work as an anesthesiologist, he was able to keep his cool in the most stressful situations. I took note and reminded myself about the lessons learned from my home burglary: tuck away the emotions for another time; they won't serve you well when the emergency strikes. My heart rate instantly slowed, and I realized this stress was the price of my entry into the high-stakes game of cave exploration.

Paul calmly emerged from the mucky water, looking relaxed in spite of the danger he had been in. I couldn't hide my relief as he crawled up the rocks. I threw my arms around his shivering body. Yet, other than being bitterly cold, he seemed completely unruffled by the experience. Of course, he had missed all the action in the gorge.

"You have no idea what happened while you were gone!" I said.

"Well, I lost some visibility, that's for sure," he replied calmly.

"I was pretty worried," I choked out, tears welling up in my eyes.

"Hey, nothing's going to happen to me."

I knew, though—and surely Paul knew as well—that we were standing on the edge of a deep chasm where death always lurked.

BACK AT CAMP, the project continued. The rest of the team worked on new climbing routes and surveying dry caves, while I supported Paul and now Noel in their dives. Mud was still flowing into the cave, so we constructed diversion dams and rocky blockades in the river basin, hoping it would improve visibility. Despite our Herculean efforts, it didn't, and after a series of dives, Noel felt he had reached his limit of comfort and experience and rejoined the climbing team on their work finding new leads, or routes, that might reveal previously unseen entrances into our cave system. I saw this as my opportunity.

I was desperate to get in the water. After more than a week in the canyon, I was getting comfortable with the magnitude of the dives I was watching others do. My own gear was still on top of the mountain, but I knew I could fit Noel's equipment and wanted to give it a shot. My relationship with Paul had also taken a turn. The closeness of shared adversity had opened the door for deeper conversations about life and future plans together, and he surprised me by confessing that he loved me. I was smitten, but not ready to verbalize my own feelings for him. I was still trying to figure out the direction for my life, and expressing love to Paul suggested a permanence that I wasn't ready to take on.

That evening, around a roaring campfire, I blurted out, "I want to dive in Noel's place." Paul's eyebrows went up, but then he smiled. "I'm okay with that," he replied.

But expeditions are democracies that require consensus, and a single individual cannot assume risk on his or her own. A risk taken for one is a risk for all. The Grim Reaper was never far away. A broken ankle one mile back in a cave could mean death: it might take too long for a rescue team to reach the diver before hypothermia set in. When a diver dies, they don't just leave behind a broken family. They require their rescuers to assume exceptional personal risk to recover their body. They leave an entire wounded community bereft. A death on an expedition can ruin reputations and end relationships with sponsors. "The Most Dangerous Sport Claims Another Life," the headlines read. Caves are sealed off. Governments enact bylaws, and people shake their heads about the futility of the sport, the waste of a life. If I was going to jump into this world of exploration, I had to be ready to take on all that responsibility and keep all those people front of mind instead of doing something stupid or self-aggrandizing. So we had to hear what the rest of the team thought.

Bill laid it on the line: "What makes you think you can succeed where Noel doesn't want to continue? How are you going to manage in the bad visibility? You might be lucky to see two inches in front of you in that crap. Convince me that you can do this, Jill, because I won't—I can't—carry another friend out of this cave."

I took a deep breath. "Bill, I want to give it a try. When I learned to dive in Canada, I couldn't even see my instructor in the murk. I just waited for her to come around and ask me to perform my skills. Dirty water and I are friends." I knew I was being naive. At this point in my career, there were more things that could kill me down there than I was aware.

Bill wasn't easily convinced, so the entire team spent a long night around the fire, sharing stories and digging deep into our personal motivations for diving and exploration. I told them about my brush with the burglar and how it had taught me a new response to fear. I told them about times I had aborted dives because things didn't feel right, and I shared that I thought it was far safer for two people to dive together than one alone.

The next morning, having been given a green light to dive, Paul and I took a stroll to look for other entrances to the cave and discuss logistics. We needed to make a dive plan that we were both comfortable with. We trekked through the bush and climbed over a ridge, discovering a neat little cavern teeming with life. Clumps of bats broke away from the ceiling and flapped around our heads in the entryway, their wings stirring up the cool air from the cave. Piles of guano on the floor created a pungent cushy carpet that was covered with weird centipedes, beetles, and spiders. The strange portal would be a Halloween nightmare for some, but the rush of cool air was all I needed to improve my sweaty discomfort.

We decided to stage some extra gas—advance and deposit supplementary scuba tanks to permit further cave penetration— and to drop some tanks filled with an 86 percent helium-oxygen mix in case we encountered deep tunnels. With passages already plunging to two hundred feet, this precious supply of heliox could help stave off the rapturous narcosis when diving below one hundred feet. It would help keep our thoughts straight.

Where Paul had left off his exploration the day before, the cave appeared to decline steeply in a "booming borehole" passage. It looked like a cave diver's utopia, but the increasing depth escalated our risk. A longer, deeper dive would require more gas and a slow measured ascent before we surfaced. There was significant water flow, a pebbled stone floor, and a large descending passage heading out of sight. The underground river was flowing strong, and the prospects for new exploration were great.

We talked through every aspect of the planned dive and all the possible emergency scenarios, and then we headed back toward camp to pitch the plan to the rest of the team. I was elated that we were working together and in agreement on the best course of action.

"We'll get through the worst visibility," Paul said, "then climb over a small waterfall into a pool on the other side. We'll drop back in the water with our extra tanks and double-check the equipment before we go under."

We climbed down a short, steep drop using a small hand line. Paul went first and landed on the leafy jungle floor, then stepped aside as I worked my way down the bluff. The rope didn't reach the ground, so I was just about to jump the last couple of feet when Paul yelled, "Stop!" He said it with such force that I froze in place. "Don't move an inch," he said. "There's a really big snake right below you."

I looked down and sure enough, four feet below my toes I saw an ominous camouflaged shape in the leaves. It was seven feet long, beige-and-brown diamond-patterned, with a reddish head. It glimmered iridescent in the sun. I hoped it would quickly move on so Paul and I could continue, but instead it rested its head on a rock while I precariously hovered above.

I had studied photos of this particular snake before leaving the States, so I knew that it was the fer-de-lance, whose venom causes certain death. To the local indigenous people, it's known as the Nine Steps, after the nine steps that victims are said to take before dropping dead. I was determined to avoid that fate and prepared to try to jump beyond it. I was hedging my bets that the snake was too heavy to leap like a cobra, but I couldn't know for sure. I lowered myself as close as possible to the ground without making any fast moves, then pushed off the cliff wall with all my strength. The snake shifted slightly in the leaves when I landed, but seemed content to keep resting. It was just another reminder that there were lots of unseen dangers in exploration.

EXPEDITIONARY TRAVEL IS unpredictable. Just when you think you have it all figured out, things get more intense. When you think it can't possibly get any hotter, the temperature rises another few degrees. Just when you think Mother Nature has doled out her worst, she finds a new curveball to throw at you. In our case, it was the arrival of a biblical-level sandstorm followed by flooding at camp.

We were justifiably worried about springtime flash-flooding but were hardly expecting what came in the middle of the night. The evening began still and hot, bugs hatching and biting everywhere, my exposed skin speckled with small red dots and rising welts. Between the shelter of the netted tent and a smoldering fire pit, I was hoping for enough protection to provide a good night's sleep, but it wasn't working. Paul was tossing in the cramped space of our previously damaged tent, dripping with sweat and swatting an endless tirade of mosquitoes. Suddenly, as I was trying to get comfortable and still myself to sleep, a gust of wind roared down the canyon, carrying with it heavy, coarse sand. There was so much of it that it passed right through the bug netting of our tent, painfully blasting our sweaty, nearly naked bodies like an emery board. I could see the fiberglass poles of our tent bending in the wind and the tent pegs lifting right out of the ground. I heard a pole snap and then we were engulfed in nylon fabric.

Paul escaped through the zippered entrance flap and gathered large rocks to hold down the tent while I splayed my body inside, trying to keep everything from blowing away like tumbleweed. We wrestled with the tent and the wind for a few minutes, but it was futile. Battered and defeated, we decided to give up on the shelter and instead lay with our bodies half-submerged in the river. I caked my exposed skin with healthy slops of mud, hoping to stay cool and uninteresting to bugs. Instead, the clay pulled my skin taut as it dried, draining the fluid from my already parched body. I wasn't having fun anymore. The discomfort, unpredictability of the weather, and stress were really getting to me. All I wanted was to be home, in my soft bed where there was no wind and no mosquitoes.

The sun eventually peeked through the ravine, but nobody in camp had slept a wink. Exhausted from battling the elements, we were not fit for diving or climbing. We aborted all exploration activities for the day and opted to rest instead. Two omens down, but I feared that bad things usually come in threes—and I was right. Shortly after breakfast, a falling tree branch rebounded off Noel's head. A head injury in a remote place could easily spell death and, as fate would have it, Noel was our team's doctor. He was nauseated and had a minor concussion, but there was little we could do other than urge him to rest. We watched him closely and woke him up every few hours to make sure his condition was not worsening.

Once Noel was resting, Paul and I dragged our gear back to the resurgence. Right away, we noticed that the cave water was as clear as we had ever seen it. Bad luck behind us, it appeared that Mother Nature was finally going to give us a break. Instead of resting, we knew we had to grab the opportunity to dive. I took my time gearing up, double-checking my tank valves, three lights, two knives, safety reels, and two computers. I slid down the rocks and let the cold water seep into my wetsuit while I clipped off the extra tanks on my harness. The sudden chill, excitement, and nerves blended together in a way that made me feel like I was about to jump on a roller coaster. I pulled up my heavy neoprene hood, then spit in my mask to prevent fogging. Step by step, I ran through my safety checklist and tested each piece of gear. Then, just when we were preparing to dive, the cave reminded us of its fickle nature. The pool started to drain, but just as quickly reversed its flow back into the cave, bringing with it the clay from the rising gorge water. I heard a faint roar outside, likely from another mudslide that had unleashed in the canyon. Realizing that we couldn't miss our moment, Paul yelled, "Dive! Now!"

In that split second, I had to make a choice: dive and try to move faster than the muddy water that was rushing into the cave, or hesitate and miss the opportunity forever. I knew that what I decided had the power to make me an explorer or get me killed. I might be celebrated as brave or remembered as a young and foolish neophyte who was literally in over her head. But I didn't have time to think or discuss it with Paul; the mudslide was coming, and Paul was going to dive with or without me. I submerged and pushed off behind him.

It was a race against uncontrollable forces, but we were determined to find the missing link of new territory that intersected Bill and Barbara's exploration from the top of the mountain. Making an important connection was always a crowning achievement but this one would include bragging rights as the world's deepest cave. Finding the physical connection between the basement and attic of the cave would consolidate decades of hard work by dozens of teams who had toiled inside the mountain.

I could still see a few feet ahead of me, so I followed Paul's blue fins through the murk. We swam furiously through the first stretch of cave, keeping just ahead of the worst visibility. The rock walls felt close, at times brushing my elbows as I swam. I reached ahead to grab a sharp crook of dolomite, then pulled myself forward to the next handhold. In a tight restriction, I snagged a D-ring on my harness in the guide line, and freeing it delayed me just long enough to be enveloped in brown clay murk. My underwater cave light was on, but it was of no use. I could hold it right in front of my mask and not see a thing. We pushed forward, gingerly grasping the nylon guide line for navigation, the comforting tautness slipping through my encircling fingers while I held my free hand in front of my face. Fending off rocks, I bumped and snagged my way onward.

Finally, after twenty-five minutes, Paul and I emerged in the calm of a dome room where we were able to surface and talk about the way forward. We submerged again, and after another three hundred feet of poor visibility, suddenly the veil fell open and I could see about thirty feet ahead of me. We had passed a pocket-sized access hole that was delivering mud into the cave. Once beyond that point, things got better, and I was able to enjoy the sculpted beauty of the hard rock walls. A few stalactites hung like chandeliers from the ceiling, while scalloped rocks told the story of eons of aggressive water flow carving divots through this hallway of rock. In other places, the geology looked like flowing water itself, beautifully curving and sinuous meanders of stone.

We rose into the place we named the Waterfall Room, arriving in the lower pool to look up at a three-foot-high cascade. The sound was deafening, but it was the majesty of this room that made me gasp in awe. Paul took off his gear and cautiously climbed to the upper reservoir. The booming white water was so loud that we could barely hear each other, but our hand signals and well-defined plan made it easy for us to get organized. I passed the extra tanks up to him and he secured them to the highest point we could reach. We needed to be sure they would survive a flood of rising water and stay in place. Once we had the four extra cylinders safely stashed, our tasks were complete. We now had additional gas stored in the cave for the next dive into unexplored territory and it was time to retreat back to the entrance through the dark water.

We made our way out through the labyrinth, the visibility so bad that the glow from our lamps couldn't help. Now that I had traveled through these tight spots a few times, I was starting to be able to visualize and anticipate each obstruction. A landmark of rock would briefly reveal itself, then retreat into blackness as the beam from my handheld light moved forward until I found the next feature. Every minute or so I would catch up to Paul and give him a "bump and go," touching his shin to signal him to move onward. While I wriggled my way through the tiny openings, Paul waited beyond each of the restrictions for me to catch up. As we squeaked through the last restriction and rose to the surface in the cavernous doorway of the cave, I now understood the extent of the flooding from the night before. I climbed out of the water and squinted my eyes in the bright sun. The river was very high and almost too dangerous to ford back to camp. It was certainly too risky to carry gear, so refilling scuba tanks would have to wait. We left the equipment behind on high ground and waded up to our chests. I could barely keep my feet under me in the high water.

Back at base camp, we discovered that ten feet of beach was now submerged. Our sad looking tent was half-immersed but staying in place thanks to the pile of rocks Paul had gathered the night before. Our belongings were soaked and filthy, but only a few items had floated away. Having experienced the ebb and flow of the river for almost two weeks now, we were no longer concerned about the muddy water; it was more of an annoyance than a real danger. When you survive a situation that you thought was risky, you tend to raise the bar higher before sounding the alarm again. And anyway, I didn't spend much time thinking about it. I was high on the euphoria of a successful dive. More people had been to the moon than had seen the place I had just visited, and knowing that filled me with confidence and pride. I knew that my next effort would lead me into territory that nobody had ever explored before, again.

OVER THE FOLLOWING days, Paul and I tried to engineer ways to divert the muddy flow from entering the cave. We piled rocks and clay in makeshift berms in areas where the river water streamed into the cave. We were in a race against time to beat the rainy season, which seemed to be coming earlier than usual. The team's cavers went climbing and surveying while Paul and I retrieved, filled, and prepared our gear for an attempt to explore beyond the bitter end of our exploration line. We made another dive to cache additional scuba tanks 2500 feet into the cave. Now, with six extra tanks waiting for us, we would be able to leapfrog tank to tank even farther into new territory on our culminating dive.

On the morning of our final push into the cave, the beach was back. The night had brought a mix of blowing sand and pattering rain, but by the time pink dawn broke, it had all settled to something peaceful. The water levels had dropped in the night, and some of our lost belongings had come back into view after being buried in the river mud for days. We finished our preparations and by midday we were ready to go. I felt like a greyhound at the start line.

We sidled into the water and headed off on a six-hour round trip into the planet. Our recent survey told us that we were headed in the right direction, so my heart was soaring with a mix of anticipation and nerves. With nothing but solid rock all around me, I floated in the water. I could now see at least six feet ahead, far enough to take in the beauty of the dimpled walls. As we approached the Waterfall Room, our first staged air tanks came into view. I rehearsed the graceful retrieval in my mind, reaching with my right hand to unclip the large stainless bolt snap that secured the bottle to the guide line. The eighty-cubic-foot aluminum tank swung easily in the water as I attached the top to my chest and the lower end to a D-ring on the waistband of my harness. I swapped regulators and began breathing off the new tank, saving the air in the tanks I already carried for the exit. Paul moved ahead of me and tried to slip through a small rocky window, but now burdened with an extra cylinder, he got caught and tore an inch-long rip in his buoyancy wing. Suddenly he was streaming a trail of bubbles. I flashed my light beam back and forth to get his attention. "Turn around?" I inquired with quick hand signals. He shook his head and pointed onward. This wasn't going to stop him.

In the Waterfall Room, we carefully climbed over the cascading chute and slipped into the upper pool. As I grasped razor-edged slivers of rock, I sliced a deep cut through my glove and into the skin of my palm. I balled up my hand and held my middle fingers tight against my thumb to stem the flow of blood, but it didn't stop the blood from gushing into the blue water. I would have to deal with tending to the wound after the dive.

Once we got to the farthest point we'd reached days earlier, we tied on a fresh line reel and swam forward in the clearest water I had seen on this trip. We descended an abrupt slope, down to a depth of 150 feet, where we needed to change over to the tanks filled with the helium-oxygen mixture. My mind was as focused as it had ever been; at 180 feet of depth, there is no room for error. Each inhaled breath at that depth consumed a volume of gas that was seven times greater than it would have been on the surface. Conserving energy and keeping calm were paramount. Every tactile sensation, sound, and visual stimulus registered sharply in my head. The bubbles from my exhaled breath roared loudly along the pitched ceiling, sounding like a freight train on the tracks. The pleasure centers of my brain were dancing with life. For an explorer, the rush we experience from exploration is like a drug. Ahead of me, I could make out the details of the sloping passage. The tunnel arched twenty to thirty feet above us and widened to almost fifty feet. Euphoric, I chased Paul through a wide-open tunnel that showed no sign of stopping.

At times, the walls disappeared beyond our view, swallowed in the blackness around us, and the tunnel would plunge, forcing my pressure gauge needle to plummet with it. Every foot of depth demanded more gas to feed our hungry lungs, and looking at our dwindling tank pressures, we knew that now was the time to turn the dive and head for home. No matter how strongly the cave called us, an excursion beyond this location could take us to the point of no return.

We had amassed a significant decompression penalty and were not able to immediately surface in the Waterfall Room. We had to slowly move up the slope in stepped delays known as decompression stops, that helped us to re-acclimate to surface pressure. At 180 feet, our bodies had been subjected to 100 pounds per square inch (psi) of pressure. Just like a pressurized soda bottle will fizz if the cap is removed too quickly, bubbles could come out of our tissues if we ascended too fast. That could cause a debilitating and sometimes fatal condition called decompression illness or the bends. We rested in the darkness with the thunderous waterfall above us, the water temperature and thin heliox breathing gas leaving us shivering with cold. As badly as we wanted to get back to the warmth of camp, it was more important to let our bodies readjust than risk a case of the bends. We turned off our lights to conserve our batteries, and I almost found myself wishing for the meager glow of warmth from the 35-watt bulb. I felt Paul trembling from the cold, and I could tell he was physically tapped out. Our bodies were wasting from the rigors of the expedition—poor sleep, physical exhaustion, insect bites, blisters, bruises, and the constant dampness. I pulled a granola bar from my pouch and broke it in half to share it with Paul. The sweetness tingled through my body in a rush. We huddled together under a rumpled metallic space blanket I had brought along and, pressing our lips to each other's ears so we could hear above the rushing falls, we talked about plans for future exploration together and what would come next in our lives. There was an incomparable intimacy and satisfaction in that moment with Paul. We had just mapped a place as foreign as the dark side of the moon. No matter where we went in life, we would always have this shared experience.

Finally, in the late evening, we popped our heads up in the entry cavern to find Bill Stone crouched by the water like a mother hen tending her flock. I don't know how long he had been waiting for us, but I could see how relieved he was to see us. Before Paul or I could say anything, Bill congratulated us. "Mission accomplished! You've given me a gold-medal day." He didn't know whether we had found a new cave but was celebrating because the most successful dive we could have given him was one that brought us home safe.

With Bill's words, everything in my life changed, and an imaginary barrier seemed to disappear. Yes, I had been born an explorer, just like every other being on earth, but now I finally felt like one.

Although our hope of finding and mapping a link to passages explored from the top of the mountain wasn't fulfilled, we had still accomplished something significant: the cave we had just dived in was now marked at 5525 vertical feet from the entrance high above us in the mountains. Nothing in the world was deeper. But it is not good enough to assume that one cave passage connects with another. Even a trace of dye poured in the upper cave and detected in the water at the resurgence at the base of the mountain was not adequate. Although we were certain the upper and lower routes were part of the same cave, a world record can only be claimed if explorers pass through the connection and map the route completely. Although we had extended the depth of the Huautla resurgence, we still had not found the missing link. Making this the deepest cave in the world would have to wait for another year, or perhaps another decade. Caves are cunning. Just when you think you are close to success, a wrench is thrown in the works that propels you in another direction.

THE LONGEST

1995

THE CAVE-DIVING COMMUNITY is represented by a rich montage of people with disparate goals. Some are casual vacationers, others are in it for life, and still others are voraciously checking off a list of feats for their trophy wall. But everyone has destinations in mind and places to go, and that is where a little competition can creep into the sport, like going beyond the end of a previously surveyed cave with bigger dives and more equipment. Every so often that competitiveness leads to cliques and clubs that claim bragging rights on certain caves and coveted gear sponsorships. Cave divers, including the great Sheck Exley, focus on records. Besides tracking accident statistics for the community, Sheck also kept track of record-breaking dives. He documented the details of the lengthiest sea cave, the longest sump, the greatest traverse between two caves, and so many other records. So when he died in 1994, while attempting the first cave dive beyond a depth of one thousand feet, people were shocked, but few were surprised. It was a shock that an icon of the sport was no longer with us, but it was unsurprising, given the extreme physiological nature of his endeavor.

In the two years following Sheck's death, the community seemed to lose its taste for depth records. Instead, everyone became more interested in measuring horizontal distances rather than vertical ones. Instead of plunging to the bottom in vertical caves, the community became more engrossed in exploring the most distant branches of some of the longest lateral cave systems in the world. As in Huaulta, connecting one cave to another was another way to land yourself or your team on the record list. These projects gave the community a new sense of purpose. Exploratory efforts filled with hardship and personal sacrifice ultimately bind everything and everyone—explorers, sponsors, and spirit—together in something we call an expedition.

In the months that followed our return from Huautla in 1995, Paul and I fell into the relationship we had been dancing around the edges of for months. We also fell into the frenzy of running his dive shop in Florida. Summer arrived, and so did every piece of diving apparatus within a fifty-mile radius. Tanks, regulators, buoyancy devices—everyone wanted them serviced for the summer season, and parents dropped off young scuba campers to participate in the least expensive and most adventurous babysitting service in town. I jumped right in with Paul, working with kids from nine till five, conducting adult classes in the evenings, and repairing regulators until we were blind with exhaustion.

I helped out with full devotion even though I had no formal residency status in the United States. I could help, but I couldn't get paid. Even though my bank account was shrinking, my relationship with Paul blossomed in the way two people who have experienced combat are forever connected. We were spending twenty-four hours a day together. Whether we were diving, working, or cuddling in bed, we were inseparable. I was happily satisfied in the intensity of doing things I loved—yet I quickly realized that running a dive shop was a big job. We were struggling to find a balance. How had Paul kept up without my extra set of hands? He had other staff, but there was more to do than all of us could manage. Then there was the fact that working in a dive shop did not come close to fulfilling my career aspirations. But how could I be away from him? The inseparable bond that germinated in Huautla had taken on a life of its own.

In 1996, realizing that the clock was ticking and my most recent six-month tourist visa would soon expire, Paul proposed and rushed me to the modest altar of the Pasco County, Florida, County Clerk's Office. We were so busy at the shop that we didn't have the time to plan a proper wedding. Paul wasn't a morning guy, but we got out of bed early, said our vows to the justice of the peace, and were back to open the scuba shop at 9 a.m. We would make time for celebration and a honeymoon later, and another, more traditional wedding in Canada would follow someday.

My "honeymoon" played out on a cave-diving project in the Riviera Maya region of Mexico. To be factually correct, tourism authorities had not yet come up with the moniker "Riviera Maya"; the region was yet to be discovered by tourists. It was an unassuming, sleepy Yucatán coastline, where the white beaches invited nesting turtles and the jungles were filled with crystalline pools of turquoise water leading to fantastic, beautiful caves.

The village of Playa del Carmen was accessible only by a single dirt road. The small town of Tulum was, at the time, a dusty, barely navigable stretch of a few shops with open sewers. Cancún was the only hotel strip, attracting boozy beachgoers. We could freely camp on the virgin beach near Akumal and never see a light in any direction. The meteor showers danced overhead like silver tinsel across the velvety night sky. We could chop into the jungle and find new blue holes—cave entrances—in the ground. We could approach a local Maya and ask, "*Dónde está la cenote?*" and be led to cerulean treasures in the woods, the watery doorways to Xibalba, the mythic Maya underworld.

Our honeymoon didn't fit anyone else's idea of a typical post-wedding holiday, but it was perfect for us. We took our time driving from Florida, camping on wild beaches, exploring Maya ruins, and hiking in the avian cacophony of the lush Chiapas jungle. We picnicked on mouthwatering arepas and fruit from colorful roadside stands. We ate savory chicken barbecued on charcoal fires and tacos with pickled onions and green salsa. We cuddled in the rooftop canopy of our old VW van and talked about the places that we had always dreamed about visiting.

With my star rising from my recent exploration work, I was now co-leading a formidable project with Mexican residents Steve Gerrard and Buddy Quattlebaum. I had recognized that expeditions rarely got dropped in your lap like Huautla had been for me. Instead of waiting for a leadership opportunity, I created one. I had raised sponsorship money for and helped to organize this new project, the Ejido Jacinto Pat Expedition. Rather than exploring mountainous vertical caves like at Huautla, we were attempting to survey enough new territory to have the Dos Ojos cave system declared as the world's longest underwater cave. For me, the opportunity would also include taking the first underwater photos from these remote places.

The three of us invited thirty-five explorers to work on a concentrated push to lay as much new line in undiscovered caves as was possible over a few short weeks. We had several smaller caves to link into Sistema Dos Ojos—connections, as we call them. We established remote sites in the jungle and targeted additional small, restricted caves that we thought might lead underground all the way to the coast.

While most of the rest of the expedition team—divers, cartographers, and volunteer porters—stayed in air-conditioned beach condos between dives, my little team of Brian Kakuk, Gary Lemme, Paul, and myself preferred to camp in the jungle and work in the remote regions at the farthest western extent of the Dos Ojos system. By being completely immersed in the experience, we could get twice as much work done. That meant that the rest of my honeymoon was spent camping in a mildewy cave called Macco's Marvels, or M1, within arm's reach of my favorite male diving colleagues, all three of them. At that point in my life, titillating cave exploration was far more exciting than newlywed sex.

Named after a local Maya porter, Macco's Marvels comprised a series of depressions deep in the undisturbed jungle. High in the canopy, motmot birds caroled their hooting song while displaying colorful clock-pendulum tails perfectly groomed to attract a mate. An occasional fox crept through the bush hoping for a meal, but the most significant population counts belonged to mosquitoes, ticks, chiggers, and bats. I had learned to love the bats, since they could each eat a thousand mosquitoes every hour. Yucatán is home to more than fifty-five species of bats, and seventeen of those roosted in these caves. At dusk, it is not uncommon to watch a swarm of tens of thousands of bats leaving the cave ceiling to forage.

Like the bats, I knew that the cave was the best place to set up camp. The jungle ground was composed of sharp-edged karst limestone under a veneer of rotting leaves. My choice was either a hammock tied between trees in the rain, and the inevitable merciless mosquitoes, or living with bats, spiders, and scorpions. I chose the latter. Under the rocky cover, I could sleep in the equivalent of air-conditioned comfort and be away from predators. If I picked my spot right, I could keep clear of drips of rain soaking through the ceiling and still be able to sleep without a net. The floor of the cave was adorned with boulders interspersed with "cave pearls," mineral pebbles the size of ball bearings, that settled around the shape of your body. Not a typical honeymoon suite, but better than the alternative.

Buddy Quattlebaum, who had a seemingly clairvoyant ability to previsualize unexplored places, suggested to Paul and me that in his experience, cave entrances were rarely separated by more than a mile of underwater passages. That meant that every six thousand feet of tunnel or less revealed a new doorway to the sky. Buddy instructed Paul to launch a solo dive in the hope that it might be a successful one-way trip. If Paul could find daylight within a third of his scuba tanks' volume, then he would surface and holler for us in the jungle. We would begin listening for his jungle yelps in ninety minutes and search for two additional hours. We'd stop every few minutes to holler for him and listen. Once we found him, we'd establish a new base of operation. If we couldn't find him from his calls, he would do his best to mark the new site and swim back to base camp through the cave. It was not a very sophisticated plan, but it was workable.

Paul began his dive from M1 in search of a western cenote that would lead the team deeper into the source water of Sistema Dos Ojos. He left the surface on a small blue and lime-green propeller-driven, underwater scooter and zipped toward the end of our line while Buddy and I sharpened the machetes and gave him a head start.

The air hung oppressively in the thick jungle. I felt heavy and slow in the humidity, but soon it was time to summon my energy reserves and begin the work of finding Paul in the bush. We set off in a northwest direction, knowing that the cave water was coming from higher elevations inland. In front of me, Quattlebaum wore a threadbare pair of denim cutoffs that barely hung on to his bony hips. The exposed skin of his back tightly stretched over his protruding ribs. He was a walking medical specimen displaying every bone in the human form. But his gaunt features belied his strength as he swung the razor-edge machete through the brush. It was as if the Scarecrow from *The Wizard of Oz* was leading Dorothy through the tropical wilderness. Except this Scarecrow had enough brains for two people. And he wielded a mean machete.

As we trekked west, I cleared small saplings and tied orange tape on the dripping chechem, or poisonwood, trees that oozed a skin-dissolving alkaloid sap. Just a few drops on your skin and your arm would begin to look like thick cheese pizza, followed by three weeks of a maddening, sleep-depriving itch that made poison ivy seem like a pleasure.

Our dive watches chimed that it had been ninety minutes since Paul descended out of sight.

"Ayaiiii-aiiiiiii. Wooop, wooop?" Quattlebaum the Warrior called.

A thousand birds, monkeys, and other animals returned with chirps, eeps, bellows, and barks, but nothing from Paul. We continued west on the same general path, led by Buddy's wilderness instincts to where we might possibly contact my new husband.

I began to doubt the effort we were making. It seemed like a wild goose chase, hoping to find a man who would emerge from a random hole in the ground in the nearly impenetrable bush. Our narrow two-hour window made success seem even more unlikely. The jungle was deafening with the calls of birds and invisible monkeys. Would we even hear Paul yell out? We were guessing that there would be a cenote in the middle of a dense jungle canopy, and we were trekking with machetes for over a mile without a compass or any other navigational tool beyond the sun. I was full of doubts. Then, suddenly—whoompf. My world collapsed beneath me.

Confused and startled, I found myself with my right leg completely punched through the forest floor. My bent left knee was the only thing keeping me from falling right through to Xibalba, the underworld. I hoped that I hadn't broken a bone—rescue from this remote place would be difficult. Precariously, I extricated myself with a hand up from Buddy.

My army-surplus pants were torn, revealing a number of bleeding cuts. I stepped forward cautiously, testing my wobbly, beat-up leg, and brushed off the untidiness of leaves and dirt, avoiding chechem branches. I looked down to examine what I'd fallen into, and it was then that I noticed the uneven ground, honeycombed with small holes. I remembered how, a few days earlier, I'd been swimming through a web of tree roots that descended from the ceiling of the caves that I was photographing. Buddy and I were walking on a thin veneer where all the trees were actively drilling through the rock to find water. I would need to be better at reading this jungle floor or risk falling through it again.

Nearing the two-hour cutoff point, I was getting more worried about not finding Paul. Dusk was approaching, and we would soon need to give up the hunt.

"I don't understand . . ." Buddy began.

Then we heard it, somewhat loud: "This will never work!"

"Paul, we're here! Keep hollering!" I answered. Buddy and I followed Paul's voice, and within ten minutes we had veered in his direction and found him well placed twenty-five feet up a large ceiba tree overlooking a narrow chimney-like rock shaft leading to a cave below. This tree of life to the Maya is the most sacred of the jungle. Paul had scaled the thorny trunk in his neoprene diving suit to reach the canopy that indigenous people believed connected the celestial skies with the underworld. For us, the new opening would mean a new base camp and the chance to explore even farther westward from this hole Paul named Conch Hope, a tribute to a cave in the Bahamas he had explored with our teammate and dear friend, Brian Kakuk.

"Do you want to hike back with us?" I asked.

Paul climbed down the tree and scanned us from head to toe. We were bleeding and filthy.

"Thanks, no, I'll swim."

FOR THE NEXT eight days, Paul, Brian, and I dived out of our new base at Conch Hope, leaving as a group each day but separating to chase our own fresh leads on underwater explorations into new territory. The cave branched considerably, and the side passages were small and crumbly, best suited for solo diving. My confidence was growing quickly, and I was giddy every time I reloaded my empty exploration reel with new line or passed on my survey notes to our chief cartographer. The group was tallying up miles of new cave passages, but I also felt that I was making tremendous progress on my own. It was perhaps the first time I really felt I had earned my stripes as an explorer, without necessarily needing to dive with Paul.

On one dive, I made a solo excursion into a very snug tunnel not much wider than the space under a kitchen table. The brittle white rock of the cave passage made it difficult to find solid places to secure the line that I would use to survey back out, but I still felt quite satisfied when I paused to install a directional marker that was inscribed with my initials. But then my reel stalled—I was out of line. Even though I had plenty of gas to continue, I had to turn around. But with time on my side, I took a moment to savor my surroundings. The water ahead of me was still flowing clearly from an unknown source, so I let the mouthpiece loosen in my lips and slowly the water flowed in around the edges of my mouth. It tasted perfect and sweet as I swished it around for one refreshing gulp before I needed to clear the mouthpiece and breathe again. I closed my eyes and felt the drift of water tickling across the skin of my face. It felt like I was inside the water source that fed the body of the planet.

But it was time to go. I spun around to retrace my route. The visibility was now near zero, so I moved carefully along the line, counting the knots that marked every ten feet of passage. Every time the line changed direction, I checked my compass and recorded my direction on my underwater slate. I noted the depth and tried to sketch the shape of the passages, which were looking more and more like a giant web of veins and arteries.

After the dive, I reviewed my notes and measurements with pride. I had placed more than a thousand feet of new line in the cave—the longest solo exploration I had ever completed—and when I plotted the numbers, everything looked good. One small error in a cave survey can throw everything off. When your work requires precision, an errant compass reading or illegible measurement can render the entire dive's effort worthless. In this case, I had a new personal best in exploration and accuracy. It felt incredibly empowering.

WE WERE PRETTY much on our own for weeks, until one morning a solo hiker appeared, joining Paul and me at the opening of the cave. "Fancy meeting you here!" he called out.

It was Michael Menduno, the editor of *aquaCorps*, my favorite dive magazine. I knew Michael was writing an article about our project, but I never expected him to trek all the way to this far corner of the map to actually see us working.

Paul and I were wrapping things up and were preparing to make a one-mile horizontal return swim to clean up all the extra gear from our far-side base of operations. With six tanks each, plus scooters, gear bags, and camera equipment, we reasoned that everything that could be swum back was one less thing to carry through the precarious, leg-trapping jungle.

By the time we left the Riviera Maya, we had cast a shadow over all other caves on the planet. The Ejido Jacinto Pat was now home to the longest network of surveyed underwater tunnels in in the world—and that was because of us, and my contributions. Though I knew the record would not hold for long, it still gave me an incredible feeling of accomplishment. That night, as we celebrated at a rooftop party overlooking a beach at Puerto Aventuras, I felt I had truly found my place in the cave-diving community. The team of thirty-five explorers from all over the world posed for a group shot in matching T-shirts, proudly grinning over our collective success.

Two months later, I received a large brown envelope in the mail. I ripped it open to find a magazine with Menduno's article, titled "Staking Out the World's Longest Underwater Cave." There was a full-page photo I had taken during our expedition, with my name in the caption. With it had come my first paycheck for a cave-diving shot. In the first paragraph, describing our meeting in the bush, there was my name again. In his closing paragraph, Michael wrote, "Cave divers that enlisted in the Ejido Jacinto Pat Expedition have laid and surveyed over 26 kilometers [85,300 feet] of line since January, and in the process connected more than 56 km [184,000 feet] of underwater passage, making 'Sistema Ejido Jacinto Pat' the longest underwater cave system in the world." I was ecstatic and hugged the magazine to my chest. The record wasn't news to me, and it was only a footnote in cave-diving history, but seeing it in print made everything more real. The outside affirmation seemed surreal. Reading my name in the story and captions clearly indicated that I was an explorer and underwater photographer. Slowly, I was starting to believe it myself.

My co-leaders contacted *The Guinness Book of World Records* to report that we had just eclipsed the longest underwater cave exploration in the world. If only for a passing moment, we could celebrate our success and safety.

What was equally important to me was to show the world that the delicate limestone cave systems and the enormous freshwater reservoirs of Yucatán might all be interconnected. That meant they could soon be under threat from rapid urban development. The dirt roads we had followed when I first visited would soon become six-lane highways. An exploding population along the Riviera Maya, with massive resorts and golf courses and the growing infrastructure of tourism, would introduce pollutants into the very places people were traveling to enjoy. The possibilities for irreversible damage were frightening. And as I moved on in my career, I would use my photography and writing abilities to tell the story that I saw unfolding all over the world. As a witness to the changes occurring in the earth's water systems, I would go on to become a strong voice for protection and water advocacy.

OUR WORLD RECORD was a fleeting one, but it did not take away the pride I felt in my part of cave-diving history. It was the perfect time to be a cave-diving explorer. Caves were being surveyed and connected at a frantic rate, with explorers staking their claim on what would become the most prolific cave-diving paradise on earth. We were disrupting the norms, developing the techniques and building the technology that would exceed all previous expeditionary efforts. Rebreathers, long-range underwater scooters, and better exposure protection were opening the frontier. Every new hole promised the equivalent of being the first to summit one of the world's great mountain peaks. But as the frontier expanded, some territory was disputed. When cave divers trespassed without a landowner's consent, the "Mexican Cave Wars" found rival explorers meeting on opposite ends of a gun barrel. The "sneak dives" were commonplace, but when word got out, some divers were willing to defend their private access with potentially lethal force. Secrecy and subterfuge were fair game, and new discoveries were boasted on newly evolving internet forums and electronic bulletin boards.

For a brief time, I felt legitimized by success and a sense of mission, but internally, I harbored doubts and hesitation. A growing thread of macho politics was brewing in my male-dominated sport. People often labeled me as Paul's sidekick, or explorer-partner by marriage. People gravitated to him at dive sites to congratulate him on our project, rarely casting me a sideways glance. We were both explorers in the mission, but I had also done months of work and fundraising to make the project happen. I should have been content with the knowledge that I was an equal partner in our dives and the co-organizer of the project, but those public omissions made me angry and filled me with doubt. I couldn't bear the same criticisms when they spilled onto the rumor mill of the internet. I read posts that referred to me as Paul's "latest girlfriend." I took those comments as an accusation that Paul was letting me dive with him only because I was willing to sleep with him.

The words stung me deeply, but it was more painful that Paul couldn't recognize how much they hurt me. I wanted my husband to defend me, to point out my merits and achievements, but I also wondered if he felt I was encroaching on his territory as an explorer. I needed to know that my husband and the community I was giving so much to had my back, but every time I considered opening my mouth about it, I felt boastful and stopped myself. Eventually I reached out to the few other women technical divers around the world, and they helped me understand that I was not alone in my experiences.

The issue boiled to the surface months later in an argument with my expedition co-leader, Paul's best friend, Steve Gerrard. Paul and I were on a hired photo trip, shooting marketing images of a local cave. Over dinner at Steve's apartment, I raised a delicate topic. He had been removing labeled navigational markers from the caves in the area, replacing them with bright orange ones that were etched with the name of the business he worked for. Traditionally, line arrows bear the name of the explorer. They are used for personal navigation cues on a dive and also to mark the extent of an exploration line that was laid by the bearer. Replacing them with what was in all respects advertising made me angry. Not only did I love seeing historic names like Sheck Exley on a line arrow, but I had placed my very own markers in a cave in hard-won exploration efforts. Commercializing our safety markers seemed like sacrilege, and I was not keen on seeing Sheck's or my names removed from caves. When I explained this to Steve, he claimed his new arrows were bolder and brighter and therefore safer. The argument grew heated, and he finally yelled at me, "Are you so egotistical that you need to see your name on a marquee in the cave?"

Steve was an elder statesman in the community, a veteran explorer. But I knew he was wrong and I couldn't let it go. "It's history, Steve, and I'm not the only one that feels this way!" It takes a lot to make me yell, and I immediately began trembling and felt nauseated. More than that, though, was that my labeled arrows were tangible evidence of years of hard work and great risk. They weren't trivial to me. They embodied all the sacrifices that I had made to get to this point in my life.

I looked to Paul for his support, but he wasn't willing to argue with one of his oldest friends. Steve and I shouted at each other for a few more minutes. When he laughed at my anger, I told Paul that I wanted to leave, that the way Steve was addressing me was disrespectful and sexist. If his idol Sheck Exley had been making the same argument, I doubt Steve would have been as combative.

There are few things I hate more than being perceived as someone who has not earned their accolades. Integrity is very important to me, and I don't celebrate my success very often. I work extremely hard for it. And yet, like many women, I am often tormented with self-doubt, feeling inadequate about my accomplishments. So after the incident with Steve, I bore down with even more determination. I wanted to prove without a doubt that I had earned everything I accomplished in cave diving. I tried to ignore the internet trolls and worked toward becoming a cave-diving instructor myself. I interned with seasoned professionals and prepared for an Instructor Institute, an examination weekend at which several cave-diving instructors run you through a gauntlet of tests and skill demonstrations. I tried to ignore comments like "We don't need any women cave-diving instructors" and "She's getting too big for her britches," the words of fellow cave instructors reported back to me from friends. I chose to ignore the sexist remarks and instead quietly prove my capability beyond a shadow of anyone's doubts.

It was lonely sitting among a group of male technical diving instructors at a yearly workshop and being overlooked for a leadership role. It was sad to hear men at dive sites talk about their life partners as "the wife." I was disappointed when my Cave Instructor examination was more like a college hazing than an objective evaluation. I didn't have many female friends who understood what I was facing, other than Paul's first wife, Shannon, who was a font of support in those days. Her risk-taking days were over, having given up cave diving for motherhood, but she knew what I was experiencing. As a former cave-diving explorer herself, she knew that living in Paul's shadow could be tough. As a pioneering woman in a male-dominated endeavor, just like Shannon had been, I felt I had to perform better than the average man just to be taken seriously.

Perhaps my encounters with sexism formed part of my life mission. I wanted to be regarded as an accomplished explorer rather than as an accomplished *female* explorer. I wanted to encourage other women to fulfill their dreams despite imposed gender barriers. I wanted other women to know that difficult endeavors are possible and success is worth celebrating. I didn't want to be the lone girl on the boat.

PURPOSE

1996–1999

AFTER EXPLORING THE "longest and deepest," and after finally achieving status as a cave-diving instructor, I was struck with a profound case of what I call post-expedition blues. For years I had been striving toward a singular goal of becoming a cave explorer and photographer. Now that I was one, it was tough to find meaning in the self-checkout at the grocery store or on a highway at rush hour. After facing the raw elements of life and death in the natural world, I felt as though my daily routine lacked purpose. Every minute of my day had been framed around involvement in a mission, and suddenly I didn't have one. Instead of getting up and going for a run, I lay in bed and watched *Today* with Katie Couric. I wondered why people cared about teeth whiteners, new celebrity romances, and Mike Tyson biting Evander Holyfield's ear. Then I began doubting my own motivations. Did people wonder why I risked my life to map wet rocks? Most of my friends, men and women alike, had found their sense of purpose in having kids. They were able to channel everything into ensuring the success of their children. Although I tried to be a part-time mother to young Joe Heinerth, I had never felt that raising a family was a fit for me, but I desperately needed to give birth to something meaningful. I wanted to know that my work would make the world a better place.

While sorting through my photos for a magazine article about Huautla, I started to wonder if there was more work I could do with Bill Stone. His exploration efforts always served a greater purpose. It didn't end with conquering new places. His work supported greater scientific goals and engineering firsts. Caves were just the proving ground for his inventions. I recalled how, a year ago, we had sat around the campfire in the canyon base camp with sandflies gnawing on our inflamed red skin, brainstorming about future exploration and new frontiers. Bill was sipping a cocktail of orange Tang and grain alcohol, buzzed on the drink and also the relief that we had completed a challenging project without incident.

"How would you like to survey and map a cave you can't see?" he had asked me.

It seemed an impossible feat, but I knew Bill well enough by now to trust that there was a seed of an idea taking root. "I'm all ears," I replied.

"Imagine driving a device through a cave while it did the mapping for you. Imagine that it did not even need to see in the conventional sense. You could make a nearly perfect map, in three dimensions, and then return to drive through that virtual space like a kid controlling a joystick in a video game. If I can get $750,000 and two years' time, I can build it," he said confidently.

Recalling that conversation a year later, I knew this could be the next quest for me. This project would be a giant leap for technical cave diving. Furthermore, an accurate 3-D map could help scientists authoritatively locate drinking-water conduits beneath the surface of the earth. Water resource protection was a growing policy issue, and access to clean water was critical for everyone. I could be a part of something that would not only push me to further improve my diving capabilities, but would also give me a real sense of purpose. This project could illuminate the challenges of protecting drinking-water resources. As a cave diver with eyes in the aquifer, I could show the world how important freshwater conservation was for the future of humanity.

Paul and I had shared some of the most intense experiences of our lives during our expeditions to Mexico. Now we were trying to chart our "normal" life together as husband and wife, and it was proving to be difficult at times. We had a dive store to run and bills to pay, but we were both driven to continue exploring. We decided that it was more important to live rich experiences than to get rich. If we made financial sacrifices and kept a full-time store manager, we could both participate in Bill's new cave-mapping project, Wakulla 2, named for the cave system at Edward Ball Wakulla Springs State Park in Florida. We were prepared to miss family gatherings and social events because we were both happiest working hard in the field. We knew it was possible to find a way to juggle operating the dive store with work on Wakulla 2, but it would not be easy.

The first public announcement about the project was planned for the annual National Speleological Society–Cave Diving Section workshop, in Gainesville, Florida. We arrived at the conference with Bill, just in time to call an open meeting. We passed an announcement through the crowd and reached out personally to notable explorers who might be willing to help us imagine the techniques that would be required to explore the cave that was, as far as was known at that point, practically endless. For some people it was their chance to be a part of something historic. Others, who also wanted to dive in this special cave, reckoned that our project sounded so outrageously gritty that we were more like a circus act coming to town.

While Bill announced his vision to the crowd, I circulated a clipboard to collect the names of interested volunteers. There wasn't a person in the room who didn't want to dive at Wakulla, so the page filled up fast. The site was often referred to as cave diving's Everest, but the only way to get permission to dive there was with a special scientific permit. The cave was deep, so new exploration was going to require a radical approach, and what Bill proposed to the crowd was like suggesting a moon walk to a toddler.

I took in his pitch like a congregant listening to her favorite preacher, absorbing details about the new life-support gear and long-range scooters he would design. The dives could take us beyond the limits of what anyone had done before, the duration of the deep exposures record breaking. What that meant for our bodies was unknown. We would need exotic breathing-gas mixtures to manage the extreme depth, new life-support technology, and teamwork that could enable missions lasting over twenty hours, but only if we worked incrementally to improve our abilities and range as divers. We would need to design new equipment, act as test pilots and develop safety protocols for the audacious dives. The risks were big. Paralyzing decompression sickness, seizure-inducing oxygen exposures, and gear failure could easily lead to the death of a diver.

You could have heard a pin drop in the room when Bill Stone described the scope of the dives he was proposing. Up until this point in history, most exploration diving had been done with regular scuba tanks. Longer and deeper dives simply meant bringing and staging more scuba tanks. But now we were at the point where that string of tanks was getting ridiculous—and fragile. If even one breathing regulator on a tank malfunctioned, a team might not make it out of the cave. For a 24-hour mission, each individual diver would need 35 tanks or more, each precisely mixed with differing percentages of helium, nitrogen, and oxygen. If we added a margin for possible failures, we would need even more. It was no longer feasible. In order to achieve our goals, we would need to test and use Bill Stone's new rebreather—a heavy life-support backpack that would recycle and mix life-support gases on the fly.

Normally divers breathe a mix of air containing roughly 21 percent oxygen and the rest, mostly nitrogen, an inert gas. The inert gas serves no purpose to the human body; while our bodies metabolize the oxygen as fuel, the inert gas gets stuffed into our tissues until the water pressure can be relieved by slowly returning to the surface. If you swim up too fast or get too much inert gas in your tissues, then surfacing can be like popping the top of a Coca-Cola bottle after shaking it, with tiny bubbles rapidly coming out of solution in a gaseous fizz. These bubbles can block blood flow, cause excruciating joint pain, or tear apart tissues in micro-bruising beneath the skin. They can lodge in the spinal cord or brain rendering a diver injured, paralyzed, or dead.

But it is absolutely critical to control more than just the inert gas load. The human body can only survive in a particular range of oxygen exposures. Too little and you pass out; too much and you might have a seizure and drown. As we dive deeper, the oxygen in our breathing mix becomes more concentrated. Like stoking a fire with gasoline, the oxygen can get too rich for the body to tolerate, causing visual disturbances, ringing ears, nausea, and seizures. When a diver has an oxygen toxicity seizure, they almost always convulse and drown.

The rebreather becomes a customized breathing-gas mixing station—as well as a recycling machine—on a diver's back. The breathing hoses recapture every exhaled bubble from the diver, sending the gas through a carbon dioxide scrubber. Tiny injections of oxygen are added back into the mix to make up the molecules the diver metabolized, and helium and nitrogen percentages are constantly tweaked to net the right breathing mix for the diver's depth. Electronic controls regulate many parameters, but the diver must be able to monitor and be able to manipulate their life-support environment. If they got it even minimally wrong, they would probably die. At the time it all made sense to me, but I knew that it would mean hundreds of hours to prove that I could use the new gear competently. I was excited about the potential to go farther and deeper into a cave than ever before, but the work ahead was daunting.

While Bill took notes on a schoolroom blackboard, most people sat in slack-jawed wonder. Some offered thoughtful ideas, and others heckled. A small group walked out of the room, chuckling and loudly declaring that we were a bunch of dangerous hacks. I had no use for the pessimists or naysayers. I was completely bonded with the core team from Huautla, and I considered Bill a visionary. I trusted my instincts about his ideas and stood with my crew. I was swept into the idea that I could make these expeditionary dreams come true.

The magnitude of our plan was great. Almost everyone in that room wanted a prominent role as an exploration diver, running the line at the end of the known cave, but if this project was going to work, we needed artists to produce sponsorship materials, software engineers to work on coding the rebreather, cooks, sherpas, and mechanics to maintain the gear. We needed people with time, companies with money, manufacturers willing to collaborate, permit writers, fundraisers, instructors, trainers, and gear junkies. It would take a global community to pull it off, each of us devoting a day, a month, or a year of our time. We estimated that the development phase of this project would take two years. Not yet envisioning myself as capable of that level of advanced technical exploration diving, I first settled into a management and marketing role, bringing my artistic skills, photography, and technical background to the group. Paul and I would provide a training home for divers to learn to use the rebreathers at our shop, and I would create the materials we needed for training, fundraising, and permits. It would be a long haul, but we were committed to the dream.

The next two years were a blur of activity. Paul and I missed special occasions and sacrificed personal time. We spent our savings of over $50,000 on gear for the project. We even sacrificed our privacy while hosting a revolving door of prospective exploration divers in our home. They came in pairs from around the world to stay in our dilapidated single-wide trailer that Paul had bought in the late seventies and placed behind Scuba West. The intensity of preparing for Wakulla and keeping the dive shop on track always seemed to take priority over our relationship. Eventually the shop needed more of our attention, and Paul felt that I should be the one staying behind. I resented the assumption, since I had put so much time into organizing the project. I wanted to be married to Paul, not his dive shop, and after all my efforts at building my reputation in technical diving and photography, I was not willing to let that go.

Everything about diving seemed to take precedence over our connection as a married couple. Instead of going out for romantic dinners, we were repairing equipment until midnight. Even New Year's Eve was an occasion for diving instead of socializing with friends. It wasn't a conscious choice. It just happened. We both seemed to love diving more than anything, even more than each other. And Wakulla became a seductive addiction. The project voraciously consumed every spare moment, devouring our savings and killing our intimacy.

As the Wakulla project grew, I intensified my personal training and gained countless hours of dive experience on the rebreather while helping other explorers familiarize themselves with the gear. We organized training weeks, and I found myself shifting into a role as one of the exploration divers. I kept up my work on project organization, but by now I had gained enough time and experience to be considered one of the chosen few. It was a little frightening to find myself in a role I had not expected. I was a young woman in a mix of international macho men whose diving histories already spoke volumes. I felt that I was just discovering my full potential.

After years of training and preparation, after hurricane delays, permit meetings, and wrangling volunteers, after hosting endless dive teams plunging into our murky Scuba West dive site, we were finally prepared to demonstrate the 3-D mapping device at Wakulla Springs State Park. State officials had received several objections to our permit application from competing divers who claimed our technology was all smoke and mirrors. As a result, the state required that we prove the feasibility of our equipment and procedures. If we could demonstrate that our mapping device worked within a forty-eight-hour window, we would get a three-month permit for the following year. If not, we would be denied access and the years of preparation would come to a very frustrating end.

A fever pitch of activity preceded the demonstration at the park. The technology team traveled from Maryland, volunteers gathered, and an eager bunch of exploration and support divers began vying for a spot on the team. The success of our permit application depended on everything coming together seamlessly, so I chose to take on a key organizational role. Only two divers were needed in the water, and Paul and Larry Green were tagged for that task.

Bill Stone rolled out the torpedo-shaped mapper. "This handle," he explained, "controls the wings that will help you keep the device level or turn it in a tighter passage. Don't try to muscle it or you risk getting overexerted," he warned. Then he pointed out a series of black disks rifled around the barrel of the six-foot-long tube. "These round transmitters will fire a sonar signal in thirty-two directions, from the mapper to the wall and back to measure precise distances. Don't get between the mapper and the wall or you will be part of the measurements!" He laughed.

Brian Pease, the project engineer, demonstrated a radio beacon that would broadcast through the rock to an awaiting surface team. This group would follow the pinging sounds through the surrounding wilderness to locate the exact point above the divers in the cave. We might swim under fields, rivers, buildings, and alligator-infested swamps while the surface team would track our progress three hundred feet above us. They would plant flags in the ground and then return to set up a survey device to accurately pinpoint the location with GPS. Though we all wear GPS devices on our bodies today, broadcasting a position from inside the earth is no less remarkable now than it was on that day in 1997.

A lot of things needed to go right to guarantee getting our permit. Preparation and timing were everything, so the night before our permit test, we gathered together to make sure we all knew what needed to be done. Site manager Chris Brown's living room was overflowing with earnest expeditioners proud to be a part of something truly significant. We filled up on barbecued ribs while I started reviewing everyone's roles for the demonstration.

"Do you smell that?" somebody interrupted.

"The ribs or the sweat?" I asked.

There were snickers of recognition that few of us had even bothered to shower in the days leading up to the launch. But it wasn't the aroma of dry-suit undergarments we smelled; it was the stench of something more sinister.

Nigel Jones, our engineer, ran out to the barn to check on the mapper's battery pack that was charging on the workbench. It was on fire. The $10,000 battery had overheated and started melting down. The fire was quickly extinguished, but we could have lost everything, including Chris's house, over a simple software glitch. Worse, it was the second major gear failure and significant financial loss. A few weeks earlier, when testing the hull of the first prototype "super scooter" at the U.S. Navy's submarine test facility, the scooter housing tube imploded. The blast, caused by a simple machining error, destroyed the $32,000 device.

These failures scared everyone into the realization that no detail could be overlooked. Success, and perhaps someone's life, depended on everyone doing their job, everyone following strict protocols, and the brand-new technology working as promised. But that is not what happened.

The following day, after what seemed like a successful first dive mission, we all crowded around Bill Stone and software designer Fred Wefer to see our first mapping efforts. The anticipation had everyone in the room fired up. The stakes were high, and we were eager to see the results. "It's go time!" Bill exclaimed, then plugged in a cable and set the download process in motion. But nothing happened. A small cursor beat in patient rhythm on the blank computer screen.

It felt as if the air had suddenly been sucked out of the room. There was a collective slouch of disappointment. We didn't know what had gone wrong, but there was no data, no information from the mapper, no map. We were devastated. We had prepared for so much, but never for defeat. With now only a day left for our demonstration, could we identify the error and repeat this entire process from the beginning?

We quickly figured out that somehow the lead diver had inadvertently deactivated the mapper's computer during the dive, and there was simply no data gathered from the mission. The machine worked fine; he had simply turned off the sonar array underwater. Given the complexity of the dive, it was understandable that he had not noticed. Juggling life support and the mapper took considerable mental agility and physical strength. We had no desire or time left for blame. It was late at night, and it would take a massive effort to recharge batteries, remix gas, prep scooters, and have a support team ready to dive again by morning. We had never expected to do a second dive, but if we wanted that permit, we would have to move heaven and earth to make it happen.

We regrouped with the same energy that had got us this far. Food and rest could wait. None of us were prepared to give up on the dream that began so many years before. The fleet of white diving vans and Volkswagen campers dispersed from the parking lot like ants from a flooded mound. Volunteers spent the night carefully remixing the necessary gases, topping up the tanks with precise amounts of helium, oxygen, and nitrogen. Nigel Jones got busy charging the precious remaining battery, a job that would take the whole night. Software guru Fred Wefer, who was struggling in the final months of terminal cancer, took a needed sleep on the living room couch, while everyone else pumped up on adrenaline and caffeine to be ready for the tasks awaiting them the next morning.

I didn't sleep that night. I had been okay with organizing support divers, but now I believed there was no reason why I couldn't have been in the water too. Now that I had witnessed a mission, I knew I was just as capable of completing one. I wanted to see the inside of Wakulla and I knew I had earned my rightful place on the team.

As dawn brightened the mist hovering over the spring basin, we again assembled our gear on the beach. Tanks were analyzed, carefully labeled, then double-checked. Regulators were tested, and the mapper assembled once more. As Larry and Paul launched off the beach on the repeat mission, our hearts lodged in our throats. Just like the day before, the mechanics of the dive operation proceeded smoothly. The team descended, and we waited for the moment that I could send down divers to retrieve the mapper from Paul and Larry, who would have hours of decompression to finish before surfacing. We had less than twelve hours remaining to prove the project's viability.

Finally, Bill declared, "We have data!" We were one step closer to success. With a small hard drive in hand, Bill rushed back to Chris Brown's house, where Fred was waiting for the computer code. Fred's software would make the numbers beautiful, turning the sonar pings into visuals that would stun the world. Or so we hoped.

Fred was clearly exhausted, but literally rose to the occasion, pulling himself from horizontal to vertical to be seated in a chair. His body was wasted from months of chemo and radiation, and it was clear he was in considerable pain and that his days with us were limited. He plugged in the data while we all looked over his shoulder like nervous mother hens. But his gaunt reflection in the screen did not tell a story of victory. He looked puzzled and confused. The data were gibberish. Bill crumpled and came to rest on the couch nearby while Fred used every bit of energy he had remaining to make sense of the data. Exhaustion eventually won, and we urged him to sleep.

The house was silent as the clock ticked ever closer to morning, to the time when the park manager would declare the test a failure and deny our permit to explore Wakulla Springs. I heard Bill and Nigel whispering through the night, trying to figure out what had gone wrong. Nobody else in the compound was sleeping. We had bet our reputations and savings on this project—we wouldn't be given a third chance. I couldn't face the criticism, nor the reality of what the last two years had meant for me. The effort of this project had isolated me from my husband. We had spent our savings and let our dive shop suffer, and we might have nothing to show for it. I had bet everything on a dream that I no longer had any control over. Everything was dissolving in a random batch of numbers generated from a machine.

It was then that Fred woke up, startling everyone out of their depressive stupor with a yell of, "I know what to do!" He summoned enough strength to hammer at the keyboard. I jumped up to watch him fix the code in his software. It turned out that the mapper had been recording the measurements in the reverse order from his software. One small assumption had turned the data into meaningless nonsense.

Once the data were read correctly, the map began to pop up on the computer. It was a colorful assemblage of points in the shape of a funnel representing the wide-open spring pool leading down to the first cave restriction. Beyond that, the arching high room known as the Grand Canyon took shape before our eyes, its bell-shaped vault gracefully peaking, and beyond that a winding corridor of branching passages that ran off into the unknown. The hairs on my arms stood on end. The cloud of points filled in and I saw the cave take shape in 3-D. It worked! The sucker worked! As we all celebrated, I knew I would go there myself and be the first to explore what lay beyond the edges of the map.

As the sun rose and the end of the permit window approached, Bill raced back to Wakulla Springs to share the map with officials. This map would not only show people where the cave conduits lay but do so in perfect synchronicity with the features on the surface of the earth. For the first time in history, we could walk over the precise location of where our drinking-water conduits were located beneath our feet. With that information, we could choose future land uses that would safeguard our most valuable asset: fresh water. With Wakulla's water systems providing drinking water for the state capital and a good portion of Florida's population, this was important science.

Our exploration permit was officially granted. Both elevated and empowered by the success, I felt inspired to think big. This mission was bigger than me, and I knew it would move my life forward with a sense of purpose. The data would frame the conversation about protecting water resources, and I would be water's spokesperson.

ON DECEMBER 1, 1998, the three-month Wakulla fieldwork project opened. Paul and I were officially named by the United States Deep Caving Team as two of the eight international exploration divers. I was the lone female lead diver, and though I was nervous, I was galvanized in my knowledge that I was capable of managing whatever lay ahead.

As the sun peeked through the woods, I got up early for a run. Each breath of damp air invigorated me, and each exhalation rid my body of stress from all the preparation work. Scuba West seemed far away. I felt positive and excited about the next three months. I would cast away the distractions of the internet, dispatch any doubts about my personal capabilities, and immerse myself in the exciting discoveries ahead.

Later that day, a huge flatbed tractor-trailer full of gear—a hulking cylindrical recompression chamber, a control shack, and a steel capsule that looked better suited for the lunar surface than a Florida cave—pulled around the backside of the historic Wakulla Lodge. A massive crane arrived the following morning and skillfully lifted three bulky blue barges into the water to serve as our new floating deck. A National Geographic Television crew arrived as we were getting everything on the barge platforms aligned. Cameras and audio booms were quickly set up on the floating decks alongside our team.

I was using a rusty steel pole to help jockey the barges into position. Looking to my left, I saw my friend Jim Schlesinger, red in the face, giving it everything he had. I stepped around the front of the pole and tried to use my weight as leverage, reaching up and yanking the giant post toward me. Suddenly it gave way and the bar slammed against my forehead like a sledgehammer. I was seeing stars, but pressed my palm on the painful wound and turned to Schlesinger. I lifted my hand from my throbbing forehead and asked, "Is it bad?"

His face drained of color and he looked as though he was going to be sick. Seizing a great opportunity for a shot, the National Geographic producer turned in my direction, and I took off in a sprint for the ladies' room across the park. I knew that "if it bleeds, it leads," and any TV producer worth their paycheck was going to capitalize on the project's first blood. Not willing to be spilled all over the NGT show trailers as the diver that got herself injured even before the project got wet, I hid from the cameras and assessed my damaged forehead. Once Jim had regained his composure, he followed me into the washroom with Steri-Strips and bandages. I probably needed a few stitches, but we covered the wound with bandages and my signature headgear, a colorful cotton bandana. Thirty minutes later, I sat for my first National Geographic interview with the bloody bandage and my eye-throbbing headache hidden from everyone.

In those days, the internet was very new, but we were determined to use it to its full potential to do educational outreach and attract more volunteers. I started a project blog, but I was internet-naive. The terms "trolling," "cyberbullying," and "flamer" had not yet been fledged. When we read the nascent internet diving forums, many of us took the negative comments to heart and allowed them to sow seeds of doubt in us. I was hurt by comments that categorized our group as "clowns" and "wannabes," but when the comments started evolving beyond competition or professional envy, I was worried. Someone sent us a package of body bags with a signed handwritten note asking us to clean up after ourselves when our project was done. Online posts forecasted our failure and wished for an accident, and a vandal even ripped the safety guide lines from a section of the cave before we began our fieldwork, leaving a note that read, "To U.S. Deep Caving Team: Don't waste your time. We got it all." This act of sabotage slowed us down, forcing us to launch dives to replace all the guide lines that had been taken out.

We discovered more notes inside the cave. "IQ over 100 needed to go beyond this point," read one, and another was stuffed in the mouth of a dead catfish tied to the guide line. I was unnerved. What could possibly compel a person to wish death on us? There were a lot of risks I had to assume at Wakulla, but I never expected to be dealing with sociopaths.

One afternoon our communications system, which provided video oversight of the deep-diving team at 260 feet, went down, and we had no choice but to quickly scramble together another team of divers to check on them. Minutes later, a young man was caught in the woods after he'd cut the cable that provided the live video feed of the divers. After that particular incident, I was sick to my stomach with stress. I could understand professional jealousy of our team, but this was outright sabotage. It didn't make sense to me; we were trying to do good with our work. That was when a confidante told me, "Lead or go home. If you're attracting controversy, it's likely because you're shaking up the status quo."

I had a choice to make: I could give up or I could get stronger. It was as if I was hearing my college roommate Kim in my ear: "Get over it." I could choose to let problems immobilize me or I could move on and participate despite the distraction.

One week after setting up the infrastructure in the spring, we were ready to begin rehearsal dives, and my diving partner, Mark Meadows, and I planned an open-water simulation of our mission in the cave. One of our goals was to get a realistic feel for the fatigue that we'd encounter on dive missions that would eventually span over 20 hours. With five hours of that time represented by hard swimming, we'd have to be fit for a workout. We spent a full day assembling and checking all our gear, and on the morning of our demonstration dive, a team of safety divers helped us get everything into the water. We began to swim in what felt like endless circles in the Wakulla basin, and safety divers were dispatched to act as taskmasters. Over the course of three hours, they held up cue cards that described particular failures that we might encounter. One read, "Oxygen sensor #2 is offline." Mark and I quickly solved the problem. Then another, more serious challenge was offered by a safety diver hovering in twenty feet of water: "Jill, your scooter just died." I unclipped my backup vehicle from the D-ring behind me and turned to Mark with a signal to abort the dive. We spun around and began our swim in the opposite direction to simulate the abort. Then they hit us with a tougher scenario: a failure of three of our four scooters. I began towing Mark with the one that was still supposedly working.

These practice sessions were both mentally and physically exhausting, just as a real mission would be, but they were important for building teamwork and confidence. And they just made us more eager to get rolling, for real, in the cave. After these rehearsals, we frequently laughed along with the safety divers, who would recount that our performance looked like a child's game of Twister. With hundreds of pounds of equipment hanging off our bodies, these emergency simulations were not very graceful. But survival doesn't have to be pretty, just effective.

Each mission required a full day of gear prep, filling tanks, checking rebreathers, getting a backup plan organized and approved, and preparing our bodies for a very long day. I had a personal ritual for prepping, analyzing, and labeling gear that I staged on the beach in a particular order. I obsessively positioned every piece of equipment and had it double-checked by an independent safety officer. Once all the gear was ready, and volunteers briefed on their roles, I took a careful approach to getting dressed. Everything had to be just right. A wrinkled sock could mean painful blisters on my feet; a bunched thermal garment could restrict blood circulation and lead to the bends. First I put on an industrial layer of adult diapers, since we'd be in the water for more than twelve hours, doubling any of my previous underwater excursions. The next layer of material was wicking long underwear against my skin, two pairs of wool socks and a layer of thick thermal garments that looked and felt like a snowsuit. The outer barrier was a dry suit with heavy soled boots and a seven-millimeter-thick neoprene hood. Even at 68°F, a temperature most divers would call reasonable, we were constantly fighting a losing battle with the water that sucked the heat from our bodies.

After spending hours assembling and checking my equipment, I waddled to the entry point on the beach fifty yards away, wearing my rebreather with twin ninety-five-cubic-foot steel tanks, a suit-inflation bottle permanently mounted to the rig, and my backpack of gear, which weighed roughly two hundred pounds. If I sat down, I would not be getting up again, so I maneuvered into hip-deep water where kneeling would magically transform the weight to near-neutral buoyancy. Then, safety divers methodically delivered to us in the water each piece of accessory gear, like the mapper, scooters, tanks, and magnetic beacons. Each piece of equipment was secured to our bodies and ticked off a master safety list. This step would take so long I often had to pee even before we started the dive.

After we left the beach and descended into the cave, the time under pressure often spanned more than twenty hours—long enough that there were shift changes on the safety team, with more people arriving or going home in the course of a single dive mission. After a mission was complete, we had one to two hours of close observation followed by hours of managing and cleaning our gear.

My first dive into the cave itself was a training mission following the route we had mapped in 3-D more than a year earlier. I was doing a practice dive with a dual rebreather system that would provide even more redundancy for dives that extended beyond ten thousand feet from the entrance of the cave. It added another eighty pounds of gear on our backs but offered the peace of mind of having a second life-support option as we neared record-breaking distances into the cave.

I floated toward the depths like a giant Transformer robot with hoses and parts sticking out everywhere. As I descended, I marveled at the stadium-sized cavern that dropped steeply to a rocky restriction. From that narrow squeeze, the cave opened into a room known as the Grand Canyon.

As I entered the massive room, Jim Brown, my safety diver, hovered high above me, casting the beam of his light just in front of my view, illuminating a great thighbone of some elephantine ancestor. A woolly mammoth or mastodon once walked into this space to find water. Perhaps it was at the end of the last ice age, when the water levels were lower and there was a way for this great beast to amble down the slope of rocks to a spring pool. I could see pieces of a huge tusk and other bones nearby.

I listened to the familiar rhythmic click and hiss of the oxygen solenoid valve in my rebreather—but high above my head I could also hear a nearly continuous stream of bubbles, the normal but noisy output from Jim's standard scuba equipment. I was nearing an hour of bottom time at around 260 feet deep when Jim's light beam panned back and forth to gain my attention. I looked up and saw him illuminate his upturned thumb, indicating it was time to start heading up. I could see his hands trembling. Something was wrong.

I swam up to him. I pulled my underwater notebook from the thigh pocket of my dry suit and scrawled a quick question in pencil: "You okay?"

He grabbed my notebook and wrote, "Flooded my suit!"

Oh my god. We were in 68°F water, facing two more hours of carefully stepped decompression stops before we could break the surface, and Jim was flooded and shivering hard. But his discomfort was the least of our worries. The cold stress could easily cause the bends.

My concern for Jim grew as we eased up the slope of the cavern a few feet at a time, but there was little I could do. Our deeper decompression stops were mercifully quick, and we checked them off the list fast. He continued to swim laps around the cavern, but struggled to control his shivering. At about 150 feet of depth, the yawning opening of the cave became filled with enough light that we shut off our diving lights. It was one of the most beautiful sights I had ever seen. The clarity of the rippled palette of blue and green was like a jewel set in the frame of the black void. I could see fluffy white clouds overhanging the cypress trees on the far side of the spring hundreds of feet away.

We spent two minutes at 150 and then slid ten feet higher. Another two minutes there, then up to 130. But as we eased toward the surface, each decompression stop was longer. The forty-foot hang lasted around twenty minutes, and then we spent twenty-seven minutes at thirty feet. These last two stops felt endless to me, but for Jim, they were agony. He couldn't hold back the shakes. Sitting still, he cupped his hands over his dive light in an attempt to get warmth from the bulb. Then he decided to swim around to generate heat, so I followed him in laps around the pond, floating above a collection of mastodon bones that had been placed there for the pleasure of tourists in glass-bottomed boats. Our last stop was somewhere near an hour long, so by the time we emerged onto the beach we had been in the water for nearly four hours. Jim was shaking so badly that he could hardly get out of his suit. He rushed back to our camp and crawled into his sleeping bag. It was a cold December day, so there was no better place to warm up.

Jim had dodged a bullet and avoided the bends, but his close call left me thinking about the fact that our exploration dives were going to be considerably longer than four hours. If I suffered a suit flood from a pinhole or failed seal, even that simple complication could prove deadly. Hypothermia often leads to the bends. It is one thing to demonstrate proficiency for problems written on cue cards; solving real emergencies inside the cave is a whole different ball game. I knew that I could prevent or control many risks with planning and preparation. I could use checklists to prepare my rebreather, and carefully examine my dry suit seals for deterioration, but other unexpected issues could only be simulated and practiced for until my muscle memory and recognition skills were so well developed that they could spring into action almost without my thinking. Electronic rebreathers can fail in both subtle and spectacular ways and later in life I would learn to deal with everything from finicky sensors to exploding lithium batteries and catastrophic mechanical breakdowns.

Although I would be diving with a buddy, I had to be responsible for my own safety. If I wasn't ready to deal with problems immediately, I could get us both killed. I became even more diligent with my safety protocols and checklists and began practicing a pre-visualization of everything that could go wrong. Before a dive, I closed my eyes and walked through the obvious things that could kill me and mentally rehearsed each solution. I touched each switch and valve in emergency drill pantomimes. In my mental rehearsals, each emergency had particular sounds and even tastes. Too much oxygen tastes like a fresh spring morning. Helium sounds like a high-pitch whistle. Nitrox gas feels warm and thick. Solenoid valves click in a rhythm that should be accompanied by a hiss of gas passing into the breathing loop. My rebreather was becoming an extension of my own physiology. I knew when something sounded wrong or felt weird and I knew what to do. Those mental run-throughs helped me slip beneath the surface of the water feeling confident that I could deal with anything that came my way.

WHEN CHRISTMAS HOLIDAYS arrived, the team thinned to a strong core. Paul's ex-wife, Shannon, and their teenage son, Joe, drove up to the park to celebrate the holidays, and we had a beautiful family-style meal in the camp kitchen. I was close to Shannon and Joe, who blended right in with the exploration team. I knew the families of many of my teammates and we all tried to extend the hand of friendship to relatives left worrying at home. Toasts were offered and tears were shed. Our tight-knit expedition team that had become a precise military unit cemented a family bond. Diving missions were going well, and we were achieving measurable results. We were on top of the world, and I was comfortable in my role as an exploration diver and felt I was an important asset to the success of the entire team. I continued to run the volunteer corps, guided tours, and media efforts and was proceeding with more aggressive dives each time I was given a chance.

On the day after Christmas, Paul took Joey home to Hudson for their traditional Quebecois family gathering while I stayed on site. A few hours later, he phoned me with bad news. They had pulled into the driveway of our ramshackle trailer behind the dive shop to find the door was ajar. Inside, the trailer was practically empty. A handmade wooden jewelry box filled with family treasures was gone. The stereo and television were gone. So were small kitchen appliances, a purse, watches, some clothing, and every little knickknack. A glass candleholder Joey had given me had been stolen, and so were the Christmas presents we had wrapped for him. We had given up a lot for Wakulla. This was one more sacrifice. There wasn't an insurance company on the planet that would have protected our unsecure trailer. "Merry effing Christmas" was all I could think.

As I sobbed into the telephone, I was swept back to when I was a college kid huddled in the corner of her room awaiting a confrontation with a burglar. Robberies are intimate, and violating. Although the thief only takes stuff, your sense of security is stolen too. As much as I believed that I had processed that grief, it still held some power over me. The incident at our trailer unearthed a cascade of traumatic memories from years ago.

But the project would not wait. As the January temperatures plunged, I jumped back into the diving lineup with my close friend Brian Kakuk. We were tasked with the first deployment of a hula-hoop-sized device that pinged a signal through the rock straight up to the surface. Tracking the beacon was a bit of an art, but our "cave-radio" inventor, Brian Pease, was always ready to find the position quickly. He carried a machete along with his sunhat, mosquito net, headphones, and sensing antennae. He was often seen cut up and bloodied or blistered from poison ivy during these fast-paced tracking exercises, but never without a smile on his face.

There was a fundamental problem with the antenna system that Brian and I deployed. Two transmitters and battery packs—what we called beacons—were positioned on one dive, and then fetched on a subsequent dive to be recharged for twelve to fourteen hours, and then the whole sequence could be repeated. Dropping only one or two transmitters seemed like a colossal waste of time and resources, so Brian and I decided to try a more aggressive approach. We proposed to take a single beacon and leapfrog it through the cave to get multiple fixes on one dive. Pease was all for it, having all manners of transport, including a canoe, at his disposal. When Brian and I left the beach for our dive, he was poised in yellow headphones and hiking boots.

From the moment we triggered our "start dive," we had thirty minutes to reach our first waypoint. We parted the dense aquatic vegetation and zipped down the sloping, bouldered floor of the cavern zone into the lofty Grand Canyon. Rapidly reaching nearly three hundred feet of depth and using a breathing mix of 90 percent helium and just 10 percent oxygen, I noticed my hands were trembling. We made a hard left turn out of the Grand Canyon into B-Tunnel and were slapped by the intense water flow. I grabbed on to a protruding rock to stop for a moment and catch my breath. I noticed Brian's hands were twitchy too. I felt like I had a caffeine buzz. But then Brian yelled through his mouthpiece block, "What's going on?" At this depth, and thanks to the helium, he sounded like Daffy Duck, and I laughed. But the gas was doing something else to our bodies too. The trembling was a symptom caused by rapid compression combined with the use of helium. The effects soon ebbed, and we turned our scooters to full power to move onward.

The current was blowing like a gale, forcing us to kick as hard as we could to make forward progress. With heaving lungs, we arrived at the first drop location and leveled the transmitter coil, carefully stretching the battery pack as far from the unit as possible to protect our rebreathers from electromagnetic interference that might scramble the computers in our life-support gear. The small room we had chosen was not much more than a bulge in the tunnel, and I was worried about being too close to the coil when I turned on the transmitter. I squinted as I rocked the rubber-sheathed toggle. Lights flashed inside the clear battery pack, indicating that the coil was live and emitting an unseen—and potentially harmful—signal. Now we had to wait for the agreed twenty minutes before deactivating the transmitter. With luck, that would give Pease a chance to bushwhack and find our location before we would repeat the process again. Our plan was to provide four different locations and then swim the coil home, doubling productivity with half the gear and negating the need for a retrieval dive.

After activating the switch on the coil's battery pack, I backed as far away as I could and struck up a helium-laced conversation with Brian. Pressing my body against the wall of the cave, I watched the coil as if it was ready to jump up and kill me. Suddenly I heard a grinding sound coming from somewhere.

"Hey, can you hear that?" I asked Brian.

"What the hell is that?" he responded.

We both silently ran through our computer handsets to ensure everything was working properly in this highly charged electro-magnetic environment. The weird sound returned. Whirrr. Crunch. Pause. Whirrrrrrr. Crunch. Pause. Whirrrrrr.

Then it struck me. "Transmission!" I blurted.

Brian looked at me, puzzled.

I used a diver's hand signal to indicate "boat."

We were directly underneath the docks where the glass-bottomed tour boats were parked in the Wakulla River, and what we were hearing was the cranky transmission shifts of a vessel backing up and turning around in the docking area. It was a sound I was getting accustomed to hearing in the park, but until now, I had never heard it underwater. We were in an underground river, three hundred feet below another river, swimming in the watery sustenance of the planet, and anything that happened up there might be seen, and definitely heard, down here.

We zipped through the curving cave conduits with our scooters and got faster at deploying the coil. Hoping he had heard our last transmission, we moved farther into the cave system while above us Pease dumped his boat and was back on foot again. The tunnels wove through the tan limestone, rocks protruding from the wall as if part of a beautiful sculpture. The silt-covered floor looked soft, with not even a single mark or gouge to indicate that anyone had ever made contact with it. We completed two more radio locations and then found ourselves at the end of the B-Tunnel line, beyond which notes from internet bullies had indicated there was nothing further to explore. I pulled a line arrow from my pouch and installed it on the end of the line, one foot beyond the marker at the bitter end. Cave divers traditionally place a directional marker with their name at their maximum point of exploration. Instead, I chose to place a tag with the name of a colleague whom I admired greatly. He was my mentor and had shared important life lessons in rebreather diving. If anyone ever reached the end of this line again, they would not find my name. They would find that of Dr. Richard Pyle—a man who would not be caught dead in a cave.

I looked at my wrist computer and saw that Brian and I had a whole hour to kill before we'd have to turn for the exit. By that point, I had already more than quadrupled my longest three-hundred-foot bottom time—minutes spent at maximum depth. I looked over my shoulder to see Brian doing what he loved best, burrowing into a small hole in the wall where water was flowing out. Apparently, this section of the cave was not sewn up after all: Brian was excitedly grunting, pointing to the place where we could explore new cave. It was a black void behind a large rock, and it barely looked large enough to pass through. Getting stuck in a small space is a recognized risk, but a strong current was coming out of the hole, so our chances to lay line were good. Brian offered me the reel to go in first, but he had located our quarry, so I just yelled, "Go. Go. Go!" He unclipped his reel, tied on to the established guide line, and squirreled into the darkness, squeezing through the pinch point in a mushroom cloud of silt.

There is nothing like the thrill of delving farther into the planet to a place you know nobody else has ever seen. Brian compressed his body through the restriction, scraping the labels off the fiber-glass case of his rebreather. The tanks on his sides clanged together as he wrestled with effort. I followed him into the murky cloud and emerged from the silt hearing him loudly cursing. I looked at his hands. He was trying to free his guide line, which had become stuck in the line reel. Jamming a reel is embarrassing, but it can also take out a vital piece of safety gear—a breadcrumb trail back to safety if you can't see through the flurry of silt. He got the reel spinning freely again, I heard, "Woo-hoo!" and he was off. Ten feet later, though, he was tangled again and swearing like a navy sailor. Again he got the line moving, but he was dragging far too much gear for the job of laying line in a small space, and so I started stripping off him anything he did not need, like the magnetic coil, four extra bailout tanks, and the spare scooter, clipping each item to the guide line as we went. We would pick them up on the return trip.

I dropped my own spare gear as Brian spooled out the line into a new horizon. For thirty minutes, we swam close to full speed, hearts hammering, until we rose up in a beautiful little chamber and ran out of line. We moved up the edge of a mound of tan-colored boulders that peaked like a mountain summit in a high dome of peacock-blue water. Striped bands of alternating white, gold, and brown encircled the room, telling a history of geologic deposition that must have taken millions of years. Fossilized sea biscuits—a species of sand dollar—jutted from the wall, indicating this was an ancient seafloor; layers of limestone deposited through time, eroded again to reveal the fossilized past. I smiled as I heard my exploration brother say, "Morgan's Room," naming the pretty spot after his cherished daughter.

In a place so far from humanity's gaze, I paused to bask in the wonder of what we had just achieved. In all likelihood, no other person would ever see this spot. I might never feel this satisfied again. I wanted to spend as long as I could absorbing every visual detail and connecting it with the immense power of accomplishment.

The quiet click of my oxygen valve interrupted the silence, but also marked the limited time left for hovering over the pristine powder silt and earthy tones of stratified rock. Brian and I enjoyed one last moment of bliss, and then his eyes opened wide: "Where's my stuff!" I laughed and gave him an OK sign. He had been so focused on exploring this newfound cave that he had not even realized I had been stripping him of his bulky gear one piece at a time.

During our exit we completed a quick compass survey and picked up all the discarded gear. When we arrived at 260 feet, we waved to the camera that was linked to the control shack. It had been precisely five hours since we began our dive, and I was physically and mentally tapped out. I looked up the slope toward the little transfer capsule, our refuge, that was dangling on a weighty cable 110 feet deep. The steel Sputnik hanging in the spring would serve as our pressurized elevator, offering a safe ride to our slightly larger decompression chamber on the barge in the spring. If it worked as planned, it would get us into a dry, safe place ten hours earlier than we could without it. Ten hours in a dry chamber is far better than the same exposure underwater, where many more things could, and occasionally do, go wrong. With the risk of oxygen seizures increasing on every long mission, we could at least prevent drowning. However, if the system accidentally depressurized from a failure of a seal or porthole, whoever was inside would be killed.

It is a vision I would never forget. I felt like an astronaut moving slowly toward the lunar module. I had just been to the dark side of the moon and was returning to base. It was absolutely exhilarating. "Who gets to do this?" I thought. But we were still a long way from entering that sanctuary in the spring. For the next four hours of controlled depressurization, we slowly swam up the last 150 feet before we could strip out of our rebreathers and tanks, pass them off to safety divers, and pull our tired bodies into the submerged module.

Our plan played out as smooth as glass, and we were still energetic when we reached over our heads to pull our bodies up into the capsule after more than nine hours in the water. Brian and I hugged, and I was almost in tears from the joy of our accomplishment. With another dozen hours of decompression time ahead, there was still a long way to go before we were finished with this mission. Through the windows of the capsule, I could see the safety team swimming our rebreathers and now-empty reels up to the surface. "Ready to move?" a voice asked over the loudspeaker, startling me.

"Yes, we're sealed up and breathing on supply gas," I reported back.

Then I heard a loud hoist squeal and bang as the operator prepared to lift our elevator to the surface. With a sudden jerk, we swung like a pendulum, and I lost balance as we began to ascend.

While we peeled out of our exposure suits, the crackling voice called out, "Ninety feet . . . eighty feet . . . seventy feet." Although the capsule eased toward the surface, our bodies were still being held at roughly 65 pounds per square inch of pressure. Finally, the glaring sun broke through the porthole and the heavy pot swung sideways, dripping condensation off the walls onto my already damp dry-suit undergarments. The capsule was slid laterally along a gantry, screeching all the way to the recompression chamber, where our pod slammed up against a sealing port. With the locking ring in place, a deafening hiss of gas poured into the chamber to equalize with our bodies. When the hissing finally stopped, we were able to swing open the two-inch-thick round door and crawl into the chamber.

Rebreather computers—like airplane black boxes—were downloaded to a laptop, and a celebratory pizza was locked in to our home for the next twelve hours. I was ravenous. We were allowed to remove our oxygen masks only for five-minute breaks, so I had to eat fast. I learned that I could shovel one and a half slices per break, and would have to wait twenty-five minutes for the following air break to resume gorging. The pizza got cold but it still tasted divine.

While Brian and I decompressed, the team downloaded our rigs and went through our notebooks and survey slates. A diver wrote out a message and held it up to the window of the chamber: "Record dive!!" I mumbled through my mask, "What do you mean?" He got on the radio to announce to us that I had just gone farther into a deep cave than any woman in history. I was never a huge fan of chasing records, because that had led to the loss of great people like Sheck Exley, but I couldn't help being surprised and excited by this news, and I flushed from head to toe. I hadn't ever considered that any sort of a record was possible, and suddenly I was at the pinnacle of my sport. I had eclipsed other women in cave diving, but was also part of a very small fraternity of extreme technical cave divers. I had to savor the taste, as I knew I might never experience this feeling again. Brian quipped, "I guess that makes me the co-holder of a woman's world record!" I laughed so hard I almost choked on my pizza.

A safety diver took it upon himself to post online that the project's worst bully "just got beat by a girl." I had explored beyond the internet troll's efforts to stop me in B-Tunnel and beyond the locations of the threatening notes left in the cave. The endless bullying and cruel online postings had left me rattled. I was unhappy that someone from my team had just poked this man with a stick. But I was also insulted by the comment about getting "beat by a girl." Was I not an equal member of this team? Or was I just a novelty participant? I had hours left in the decompression habitat before I could speak to anyone about it, which I was glad for. It gave me time to cool off and let it go. I wanted to celebrate our dive, not start an argument.

MY DIVES WITH Brian Kakuk and other partners were going well, but I was finding it increasingly difficult to dive with Paul. The last big Wakulla dive we did together was on my birthday. It should have been fun for us to dive together, but instead it was incredibly stressful. Paul worried me in many ways. He was the most naturally graceful man I had ever seen in the water, but he never seemed very concerned about risk. His attitude was always, "Don't worry." Yet here we were, doing some of the most dangerous cave dives ever imagined.

I preferred tight pre-dive planning, rehearsal, and methodical preparation. Paul was a lot faster getting ready, and I always wondered if there was something he might have missed. Granted, we were alternately running back and forth to Scuba West, but I wasn't always certain his head was focused on Wakulla. On the other hand, I think he was growing frustrated that I was not at home taking care of the shop. We had been arguing more frequently. He had recently told me that he "just wanted a wife." It seemed there might not be room for two explorers in our marriage.

On my birthday dive, Paul and I set off on a mapping mission. At first I was thrilled that we were diving together. After the Christmas break-in three weeks earlier, and hard work juggling our responsibilities, we were finally on a mission together. Our job involved heading down a gargantuan main passage called A/O Tunnel. The visibility was poor in this passageway, so it would not be as visually stunning as my recent dives in B-Tunnel with Brian. I was the lead scout diver, and Paul was assigned to follow along with the cumbersome mapper. Although we were diving as a team, I had to stay far enough ahead to remain out of the sonar scan while illuminating a path for him to follow. Poor visibility made that tough. By the time I got out of the scanner's beam, I was almost beyond the range of his vision, and he mine. I could hear the hum of the mapper's propeller as he moved. The mapper was equipped with a dead man's switch, so I figured if I heard humming, he was still driving. We followed a thick golden guide line while surrounded by complete blackness. The tunnel was so vast that even with lights, we could rarely see the walls or floor or ceiling for reference.

On this day, my scooter must have had a little more pep than Paul's because I kept getting too far ahead, and he would lose track of me in the brownish water. I kept pausing to let him catch up, and one time turned back to fetch him. Each time I got ahead, though, he grew angrier. He ended up yelling through his rebreather at the top of his lungs. I couldn't understand what he was saying, but it felt like daggers to my heart. I assumed he was frustrated with me, and not with the struggle with the mapper and the visibility. It was tough to take. Whenever Brian Kakuk got upset over a tangled reel, I never once thought it had anything to do with me. When Paul yelled, I was certain whatever was happening must be my fault. I tried harder to stay in the right range, but still kept getting ahead. Dragging all the gear and staying on the line took intense concentration.

When I turned back a second time to fetch him, he was livid. The contorted high pitch of his helium-laced voice was completely unintelligible. Now underwater, in this moment of anger, all the tension between us over the last little while came flooding back to me. I felt rejected, and tears spilled into my mask. I was three hundred feet deep and taking one of the greatest risks of my life, and my emotions were getting the best of me.

At that moment, I decided this would be our last big dive together. Tears and emotions were commandeering my balance and reason. Underwater, such little incidents can get someone hurt or killed. People who are distracted can make fatal mistakes. Paul had yelled at me on other dives. When I failed to understand his modeling instructions on shallow photo dives in Mexico, I had experienced his anger. When I wanted to head up early from an offshore wreck dive, he was not pleased. But at three hundred feet down, this was much more dangerous. Here, on a dive on the physiological edge of our sport, was not a place to be angry or to feel victimized. I didn't think it was anyone's fault—we were both trying hard to do a difficult thing—but I now knew that diving with my husband was too emotional. When I told him later that I was not going to dive on any more Wakulla missions with him, he was furious.

"You don't want to dive with me?" he yelled. "You are my wife!"

But I had made up my mind. Our safety, and perhaps our marriage, depended on our diving with other partners on the team.

That revelation was certainly a part of something bigger that was brewing in our relationship. Paul and I had married as exploration partners. Was I his wife or his dive buddy? Could I be both? If we were not going to be able to dive together, I didn't think our marriage would survive. He had opened the world of exploration to me, but wasn't happy about me living in that world. What did that mean for our future together?

BEYOND THE PSYCHOLOGICAL stresses of tough dives, we all became aware of a toll on our bodies. At times we were close to exhaustion, and new physical symptoms were setting off alarm bells. We had all signed up knowing that we were essentially human guinea pigs, diving beyond the limits of what was known to diving researchers. Now we were beginning to discover what that felt like.

Sleep deprivation was a huge issue. By the time I finished a mission and climbed into my sleeping bag, I might have been awake for thirty-five hours—what most would call a full work week. Then, after three or four days' rest, I would do it all over again. The schedule took its toll on all the divers. We were completely exhausted. As well, attrition was shrinking the volunteer corps, and stress brought out frustration in some people on our team. But beyond those issues, we also discovered that oxygen exposure was taking a chunk out of our bodies.

On a dive late in the project, I began to experience bouts of nausea when I reached my decompression stop at 160 feet, several stops before I could enter the transfer capsule. I wasn't sure whether I was getting hungry, tired, or sick. Questions ran through my mind as I tried not to throw up into my breathing loop. Had I pushed it too far? Was a seizure coming on? We had totally crushed the allowable oxygen exposure limits. By the time the dive was over, I would exceed them by 400 or 500 percent. Overconsumption of oxygen at depth is like throwing too much fuel on a fire. The fuel was the oxygen in the breathing mix and the fire was in my cells. My body simply couldn't process that much oxygen without detrimental results. I needed a certain amount of oxygen to get me out of decompression, but I was walking a thin line of having too much because of the length of the dive.

When a safety diver popped down to check on me, I wrote him a note on my wrist slate to explain what was happening. I lowered the oxygen concentration in my rebreather to ward off toxicity symptoms, but doing that meant I'd have to spend additional time in the water for decompression. Oxygen helps flush away inert gas from a decompressing diver, but it can also cause a seizure in high doses. I tried other solutions, such as switching to pure breathing air on a scuba tank, but nothing helped me feel better. I grew more anxious as the nausea became more intense, and I asked the safety diver to stay close. I knew the longer I was exposed to this rich breathing mix, the more likely I was to have more severe symptoms, and that the next step could be an oxygen seizure. The sooner I could get into the capsule, the better. If I had a convulsion there, the team could help me. Underwater, I would certainly drown. I wanted to rush up the last twenty feet to the pressurized capsule, but doing so could have killed me too. I was on the knife-edge of what was humanly possible, balanced between the limits of oxygen toxicity and potential decompression sickness. I had absolutely no control over the outcome. I just hoped the safety diver at my side was ready to help if I needed him.

It was an enormous relief to enter the transfer capsule. Safety divers stripped me out of my bottles and rebreather as quickly as they could. While breathing from a long scuba hose, I reached over my head and grabbed the circular rim of the entry portal. Pulling my body upward through the hole, I pivoted and landed my bottom on the edge of the ring. I hiked my knees to my chest and yanked off my fins and mask and inhaled a fusty aroma of sweat and mildew. I wanted to throw up but was relieved to be safe again. For the next nine hours, I was unable to sleep or eat; all I could do was drink water and try not to barf. The operator gave me a couple of lengthened breaks from the oxygen mask, but I just wanted the ordeal to be over. Each air break meant an extension of the overall time in the chamber.

When I finally exited the habitat, I immediately ran into the bushes and hurled watery vomit for ten minutes. I felt better, but questions hung in the air. Was I coming up hard against my body's limits? Was it a matter of physical stamina, or did I need more recovery time between dives? I was feeling a slight burning in my chest, and after speaking with my colleague Richard Pyle via email, we decided that I was probably experiencing pulmonary oxygen toxicity, a gradually deteriorating breathing capacity caused by high doses of oxygen. Although I would fully recover from the condition in a few months, the lengthy missions were getting riskier for me.

At the post-dive meeting, I asked, "Are any of you guys feeling any of this?" I was nervous about sharing, since I understood that my confession might take me out of contention for future dives on this project. The safety committee might decide that my body was too compromised.

Others exchanged glances, as if hoping for someone else to speak first. After a moment of awkwardness, another diver revealed that his fingertips were red and hot. Others confirmed similar symptoms and strange post-dive maladies. Mark Meadows had already confessed to significant pain, especially at the sites of old injuries. My lonely confession had opened the door. With exhaustion creeping up on the team, we decided to schedule a day off so that exploration divers and volunteers could rest, heal, and regroup.

ON THAT DAY, Wes Skiles was leading the National Geographic crew that was filming our project. I watched them unload cameras and bulky lights from his big white production van. I desperately wanted to join his filming crew, but at the moment I was so exhausted I could barely speak. I went back to camp and crashed hard in a sleeping bag in the back of my van.

Wes and his crew took our day off as an opportunity to set up some lighting for the next exploration dive. Four divers planned a ten-minute dive to set up lights in the cavern. Diver Jason Richards, too focused on wrestling with a cumbersome lighting cable, breathed his pure oxygen tank all the way down to one hundred feet, which was five times deeper than pure oxygen could be safely used. Roughly eight minutes later, he lost consciousness, inverted, and plunged toward the bottom in a twitching seizure. Fellow diver Mark Long spotted Jason first, but it took all three divers to haul their blue-faced crewmate to the surface and out onto the beach. Richards was violently spasming and vomiting while first responders cut him from his brand-new dry suit.

I was just waking up from a post-dive sleep when I heard someone speed into camp and yell out an alert. Thinking that one of my friends had just died, I rushed over to the park's front gate, where I saw Jason on the ambulance stretcher, splayed out in his military-issue long johns. I was happy he was alive. Jason was a combat helicopter pilot who was known for his proficiency, and somehow he had made a near-fatal error. If it could happen to him, it could happen to anyone.

Jason spent the night in hospital and was released the next day with a clean bill of health. We helped Wes's team finish their filming, but we still needed a rest break. So, five days after Jason's close call, we had a complete expedition stand-down to reinforce procedures and clarify the way forward. I started my day catching up on computer work with Bill in one of the work trailers. But the peace was quickly shattered by Wes Skiles practically taking the door off its hinges, yelling, "911 on the beach!" Then he burst out the other door, still yelling.

Bill and I jumped to our feet and rushed down to the water's edge. A single diver was dragging a limp body in a yellow rebreather out of the water. As we ran to the water, Bill was yelling, "No! No! No!"

In 1990, Dr. Henry Kendall had been awarded the Nobel Prize in Physics with two other scientists who first described the subatomic particle known as a quark. Beyond being a brilliant mathematician and physicist, Kendall was a real adventurer. A well-traveled diver and mountaineer, he saw a bit of himself in the younger Bill Stone and became a close friend and mentor. He invested in Bill's companies and exploration work and was one of a very few "civilians" who bought one of Bill's early rebreathers. When Wakulla 2 launched, Kendall wanted to visit, and even at seventy-three, he seemed fit and eager to go for an easy dive in the cavern zone of the great spring. Bill had given the nod for one of the safety divers to escort him for a casual dive on our day off.

Later, a few people said that Kendall wasn't quite himself that morning. He had forgotten the name of a waitress he had talked to for many days in a row. He had fumbled with his dive gear. When he walked toward the water, his dive partner said, "I'm not quite ready. I'll be there in a minute."

But Kendall was kitted up and feeling overheated, so he stood in knee-deep water to put his fins on. He was inhaling from his rebreather loop but had failed to activate the oxygen supply. Standing on one leg like a pelican and struggling to put on a fin, he likely got dizzy and fell over in less than two feet of water. Unconscious in the water, he drowned. Five days earlier, Jason Richards had nearly killed himself with an overabundance of oxygen, and now Dr. Henry Kendall had unintentionally starved his brain of the necessary oxygen to stay awake.

I helped drag Kendall onto the beach with his gear still on his back. It was like trying to lug a 350-pound piece of Jell-O through wet sand. Bill and I laid him on his back right on top of the rebreather unit, ripped his mask off, and started mouth-to-mouth resuscitation, despite the fluids that oozed out of his mouth. I started CPR while a team of busy helpers peeled Henry out of his gear.

"Come on, Henry!" Bill screamed. "Breathe, man!"

"Bring some oxygen!" I shouted.

I kept pressing on his sternum and felt a soft pop, a snap of his ribs giving way to the compressions. Tears filled my eyes, and I became an automaton pressing in rhythm with Bill's attempts to breathe life into Henry. With every press, I was talking in my head: "Stay alive. Stay alive. Stay alive." I felt a shiver run up my arms and suddenly I sensed Henry all around me. It was as if he was floating over my head, looking curiously down on his body. All of a sudden it felt like there was nothing under my hands and everything above my head. Henry swirled around Bill and me like an embrace, and I was engulfed in a peaceful mistiness. I felt like Henry was telling me it was okay.

EMTs arrived and slapped leads on his chest.

"Finally!" I thought. "This is your chance. Come on, Henry!" I urged, hoping that love and encouragement might reassemble the broken physical presence before me.

The defibrillator fired with a thump, but did nothing but jolt Henry's lifeless body. They tried again.

I thought the EMTs might take over from us, but they asked us to resume CPR and radioed a supervisor. Needles and bags appeared. Leads and machines. I just kept pushing on his chest while the paramedics threw technology on Henry's cold, bare skin.

The rest is a blur. Henry was prepared for a medevac flight and taken from us, and everyone felt a brief surge of hope. Bill wanted to check Henry's dive gear to determine what had happened, but I immediately stopped him.

"Bill, stay away," I said. "You, above all, can't touch his gear. You're the designer and manufacturer, and he might be dead. We have to maintain a proper chain of custody. I'm going to film it and give it to the police."

I grabbed Wes Skiles and asked him to roll his camera. We moved the gear away from the water and Wes filmed every detail without tampering with things that could later inform us about what had happened. The problem stood out to me right away. The oxygen injection switch was in the off position. Henry's tanks might have been turned on, but no oxygen was going to the loop. Henry was typically meticulous with his checks. It was hard for me to believe he had failed to execute such a simple task, and one that should have been prevented by the electronic pre-dive checklist.

Before I could complete my review, we got the news. Henry was dead. I had been strong up until that moment, but crumpled to the ground with my face in my hands, sobbing and unable to speak. I climbed the diving tower overlooking the spring and sat alone with the wind thrumming through the Spanish moss. Sunbeams still penetrated the cerulean depths. Alligators still lounged on the shore. Nothing had changed, and everything had changed.

We later learned that Henry had been a ticking time bomb. His autopsy revealed that an undetected severe physiological issue had likely affected his mental agility that day—probably the reason he failed to activate his oxygen controller. He was dying before he even prepared his gear for his dive.

We were devastated. Bill had lost another dear friend. The world lost a brilliant adventurer and scientist. We all lost a bit of something that day, but we also grew a little wiser. When you embrace a person at the moment that their life force ebbs, you are changed. Whether you believe in a greater power or not, you cannot deny the difference.

AFTER A SAFETY stand-down and a park investigation, we were cleared to relaunch our activities for one final week of our project. It took confidence to reconcile the concerns of the state park authorities with the reality of the two accidents. It also took mental stamina to go on with the project. We had to be frank about what had happened and assure officials that we could prevent further incidents. Brian Kakuk graciously volunteered to quit diving and become full-time dive safety officer. And I discovered a louder, assertive voice that brought about an even more confident position in managing and executing dives. I spoke up about who I thought could conduct the final dives of the project and deferred some of my personal desires for more exploration in order to just get our job done. We had not saved Henry Kendall, but I had to ensure we prevented anything else from happening.

Television personality Boyd Matson was on site, eager to participate in a sequence for the National Geographic *Explorer* special about our project. He had been training as a cave diver for months in order to make a short excursion into Sally Ward Spring in the Wakulla Springs State Park. We were assembled and prepped under intense theatrical lights on the edge of the swamp but waited until the final countdown to midnight, and the resumption of our permit, before anyone stuck a toe in the water. We were not going to break any rules that could affect the last precious days of the project. The opening to Sally Ward Spring sits beneath a canopy of underwater grasses and is well known for its gargantuan alligators. This night was no exception. A dozen sets of red eyes reflected back at us from the swamp. I was told that our presence would disperse the gators, but couldn't help but think that these reptiles with walnut-sized brains might be hungry and I would provide a substantial meal.

Matson, a little nervous, said, "Boy! I guess we'll be the first people to ever dive here at midnight!" The whole cave-diving crew broke out laughing. Sally Ward was one of the state's most popular "sneak diving" sites. With permission to dive difficult to attain, many cave divers in the past had trespassed into the slough in full equipment and quietly explored the cave under cover of darkness.

As long as they were out before the park gate opened in the morning, they usually went undetected. Some went so far as to swim through a storm sewer pipe to reach the swampy entrance, but given the likelihood of meeting toothy reptiles, this probably wasn't the best idea.

We pulled off a successful dive with Matson and then took him for a dawn ride in the transfer capsule, inviting park manager Sandy Cook along for the descent. It was her first underwater experience, and she was thrilled to see this other side of her park. Wes Skiles had several other filming sequences to complete, and I offered to drive the mapper and operate a deep camera in places he was not prepared to go since hanging up his deep-diving hat after numerous episodes of decompression illness. I volunteered for every bit of film duty I could get, eager to enhance my shooting résumé. It was a great opportunity to learn from Wes. Perhaps helping him could open new doors for me.

On one of the last nights of the project, I stood by the campfire at base camp talking with Wes about Henry's death. Wes had experienced the loss of many colleagues over the years, and his empathy helped me recognize the need to grieve. He encouraged me to honor the feelings, and not to be afraid to talk about the details. For me, it wasn't just the loss of a great man. The visceral recollections of group panic on the beach, the sound of the sand and rocks beneath Henry as we dragged him out of the water, the smell of his dying body were all imprinted vividly in my mind. A similar sound or smell might suddenly jar my attention and flood haunting memories back into my head, even unrelated traumas from my past. At times, they made me feel burdened and distracted. I asked Wes if that was normal. He responded, "Normal? Jill, it's just what makes you human. Everyone experiences life's challenges. It is what we do with them that matters."

I had a lot to think about, but at that point, Wes shifted to a more pressing subject. "Jill, you have the talents I need for my business. You're an artist, organizer, and great diver, and I need a producer. What do you think?"

That was a big segue, but I confidently responded, "Great! I'll do it. What's a producer?" We both laughed.

That conversation propelled my career forward. If I worked with Wes, I would have the chance to draw on the entirety of my creative skill set. I wouldn't leave exploration work behind; I might even get to do more. If I was going to make a real career in the underwater world, I needed to be multifaceted. Writing, organizing, diving, and photography and video skills would all be important. I needed to get as comfortable in front of a camera as behind one. Specialists in niche work can always be found, but people who have a hybrid list of talents were tough to find. I had landed in the right place at the right time. Just as Bill Stone and Paul had been mentors in exploration, Wes Skiles would become a mentor in the world of underwater filming. As a friend who knew me well, he would also help me find my best and happiest self.

I ran a camera for Wes on the deep dives in the final days of our permit. I loved the additional responsibility and the opportunity to have a hand in creating the imagery that would appear on National Geographic Television. It was a thrill knowing my name would be added to the creative credits in addition to the exploration team. And in the end, the U.S. Deep Caving Team delivered a remarkable, accurate 3-D map of more than forty-two thousand feet of passages beneath Wakulla Springs State Park. The data astonished communities of engineers, geographers, and computer specialists outside of diving. Wakulla 2 transformed technical diving with its revolutionary use of rebreathers and was equally heralded in engineering and technology magazines.

My confidence and assertiveness soared in other aspects of my life. I told Paul exactly what I wanted to do with my future. I was excited about the opportunity of working with Wes, whom he had known for years. With a new confidence in my abilities,

I recognized how important it was that I pursued things that fulfilled me. I would never be happy if I gave in to society's expectations and tried to do only what pleased Paul. Having creative work to look forward to helped, and I was on a post-expedition high. Everything I wanted felt possible and positive. Paul, on the other hand, was uncertain. He liked the sound of that life too, but he had a far greater attachment to his dive shop. He was willing to see me try, but I don't think he expected me to succeed.

A week after the Wakulla 2 project wrapped up, I was on a boat off the east coast of Florida with colleagues from Bill's company, Cis-Lunar Development Labs. I was putting together marketing materials to promote their rebreathers, and we were shooting stills and video in peaceful warm water. I felt incredibly free and light. I had shed hundreds of pounds of extra gear from Wakulla and was doing much simpler dives. I enjoyed the relaxed pace and camaraderie with the other divers on board. We weren't under time pressure, so we lounged on the deck in between dives talking about the success of Wakulla rather than the loss of Henry Kendall. It was the first day I felt the stress melting away and allowing me to move on.

I was slowly preparing my equipment, basking in the enjoyment of doing exactly what I wanted. I must have had a smile as wide as the back of the boat. A fellow instructor was hanging on the tag line off the stern while his student completed his pre-dive checks. Out of the corner of my eye, I watched the student putting on his fins, careful not to lose the breathing loop from his mouth in the slight chop. There was a large splash, then he was in the water and swimming toward his instructor. I returned to my own rebreather to complete my electronic checks. A minute later, I heard a distant cry for help.

I turned to see the instructor with the blue-faced student in his arms, struggling to stay afloat in the waves well behind the boat. A moment before the student was full of life and standing on the deck, and now . . . Again, I was faced with someone close to death in the water. I snapped from complete bliss to overwhelming stress in a single heartbeat. Still wearing my pink windbreaker and U.S. Deep Caving Team ball cap, I dived off the boat and swam quickly to the instructor, blurting out, "I've done too much fucking CPR this month!"

I was still processing the attempt to save Henry Kendall. For weeks, I had felt like Henry was attached to my ankle and I was dragging him along through the world. I couldn't get my mind off him. I couldn't escape the smell of his death. I couldn't forget the feel of his cracking ribs as I pushed on his sternum. But I had thought it was over. I thought I was okay. Now it was happening all over again. The joy I'd been feeling moments earlier was gone.

I reached the instructor and rapidly towed the student to the swim platform, where other crew were crouched ready to pull him onto the boat. He was weighty with a rebreather and bailout tanks, but I managed to get his head on the swim platform and ripped off his mask. I blew two breaths into his cold mouth while I held his airway open. I was angry, feeling he had robbed me of bliss and sent my heart tumbling back into a darker place. Then, in an instant, he magically coughed back to life and breathed on his own. It had worked! Two breaths and he was alive! The team pulled him aboard, and he was soon able to sit up.

I had found the job I wanted to do, but it kept getting up-ended with terrible accidents and death. Was this the price of chasing a dream? I should have been dancing with delight that we had saved a life, but instead I felt the sobering reality of a career in technical diving. It gets very real when there is a dying man on the deck.

In the end, it turned out the diver had made a mistake similar to Henry Kendall's. He ran through his checks without watching his displays and had failed to fully activate his oxygen supply. He was lucky to be alive. Knowing how easily a person could die on rebreathers made me even more cautious. I would still be conducting deep and challenging dives, but my protocols had to be impeccable. There was no room for mistakes.

I EMERGED FROM the depths of Wakulla Springs as a world record–breaking diver, but the project had bittersweet memories. I learned that I could do far more than I ever imagined. But I'd also learned that my best was sometimes not enough for some of my male peers in the sport. Until now, I barely made a dent in the glass ceiling that discouraged so many women from pursuing a career underwater. After my success at Wakulla, I became even more determined to change that for myself, and others.

I discovered that all the strength and willpower I could muster wouldn't necessarily save a man's life. And that a woman's diving record won't change a bigoted man's hostility. But I also discovered I had choices. Career choices, relationship choices, and happiness choices. What I chose after Wakulla was to thrive in the life I could make for myself. I couldn't change the frightening or negative experiences in my past. I could not change nor be responsible for other people's actions or behaviors. Those life experiences served me in some way. They had been transformative, all helping to build the person I had become. So I did not need to regret, mourn, or fear the past.

TOP: Perhaps my first expedition. I was never happier than when I was outdoors and dirty.

BOTTOM: Close supervision from Mom at a summer cottage in Ontario, Canada.

TOP: I was proud of my Girl Guide and swimming achievements, so Mom sewed my hard-earned badges onto my sweatshirt.

BOTTOM: As a young teen, I was a competent paddler and often went on adventures with friends or my sister Jan.

TOP: One of my earliest dives, in Tobermory, Ontario. I didn't know then that ice diving would end up opening doors to future adventures in the polar regions.

BOTTOM: This might be the only work break I ever had in Cayman—a modeling gig for *Skin Diver* magazine.

TOP: Climbing up from the riverbed into the cave in Huautla, Mexico. Water gushes out of this portal like a fire hydrant during the rainy season.

BOTTOM: Shooting with my trusty old Nikonos V film camera, with Paul Heinerth at my side.

TOP: Preparing to cave dive with Paul during our honeymoon in Mexico.

BOTTOM: With cave diving becoming more popular, operators needed photos for their tourism brochures—so here I am freezing to my core in a sporty pink lycra suit.
Photo by Paul Heinerth.

OPPOSITE: I'm driving the digital wall mapper, pausing briefly for Wes Skiles to get the shot for *National Geographic*. *Photo courtesy of the U.S. Deep Caving Team Inc.*

TOP: Lugging the rebreather and two large steel bailout tanks at Wakulla Springs. I was walking with roughly 200 pounds of gear, and preparing to get in the water to add another 300 pounds to the payload.

BOTTOM: Sitting in the recompression chamber at Wakulla Springs after a twenty-two-hour mission. I was feeling a bit of nausea mixed with pride.

FOLLOWING: I tried to take this very shot of the Pit, in Mexico, on a film camera in the 1990s, but the motion blur from the long exposure ruined the photo. Returning in 2018, I finally captured the image that had been haunting me since the day I got bent.

THE PIT

2000

ONE YEAR AFTER Wakulla I heard a phrase that I had never expected to hear: "Never dive again." The physician's words made my heart drop. "I have no idea what will happen if you choose to go back in the water again."

Although I had faced the deaths of colleagues, I had not fully embraced my own mortality until I had a terrifying close call. Tiny bubbles from decompression sickness were simultaneously dissolving my body and vanquishing my faltering marriage.

When Paul and I first returned from Wakulla, we were swept into a sort of powerlessness of inertia. We worked long hours in the store and were constantly dealing with our customers' needs instead of our own. I was traveling for television work with Wes Skiles, marketing jobs for Cis-Lunar and frequently headed to North Florida's cave country to teach classes. Life was so hectic that we did little to address problems in our relationship, instead blindly fumbling along the conveyor belt of life, each doing the best we could.

But in the spring of 2000, we decided to take a break to explore a sinkhole called the Pit, in Mexico's Yucatán Peninsula. Three years earlier, we had found a deep basement cave in this system and had been keeping it secret until we could have a good crack at exploring it. It had taken us three years to assemble the technology and money necessary to get us back to the base camp deep in the jungle of the Ejido Jacinto Pat region. However, I was concerned about how some of our Wakulla dives had gone down. Diving with Paul was like putting on an old pair of comfortable shoes: they felt great when you first slipped them on, but they might not make it through a really tough day on the trail. Still, I wanted to find the expedition magic with him again. I remembered watching the wind in the Santo Domingo canyon blowing through his curly brown hair and being captivated by him. Perhaps we would recapture our love in the country where it first took root.

Once again, we thought we had an opportunity to make a potentially historic dive. On our rough sketch maps, two massive neighboring cave systems appeared to overlap, so it seemed possible to us that somewhere in the labyrinth of passages, we would find the physical link. These were the two longest caves in the world at the time. Connecting them would be remarkable. Doing so through a tunnel almost four hundred feet deep would be a complete surprise to our community.

Underwater cave exploration was expanding in the Yucatán, and the competition to be the first explorers to join two of the world's largest cave systems was getting fierce. Diving businesses along the Caribbean coast were eager to reap the rewards of being able to say they were operating in the world's longest cave. They would align themselves with one team or the other and jealously guard information that might give their team an edge. Divers were sneaking miles into the jungle to explore new openings, hoping to discover the elusive connection. Every month or so, newly found passageways and caverns were explored and the boundaries extended. Things were getting hot, and Paul and I had caught the fever.

We believed the link, if it existed, would be found at a deeper level than most explorers had ever been. Most caves along the Yucatán coast are relatively shallow, less than 80 feet in depth. We guessed the hidden passageway would be found in the older geology, some

350 to 400 feet down. Not many divers were experienced, equipped, or capable enough to explore at those depths, so we quietly returned to Mexico to explore the promising lead we had left dangling from years before. We also felt like it would be a great way to spend valuable time together and work on repairing our marriage.

Today, divers flock to the Pit by an easily accessible road, but during our exploration of the system, getting gear to the dive site required exhausting hikes, makeshift donkey carts, machete-wielding robust Maya sherpas, and a lot of determination. We had been anticipating the help of a small local support team, but upon our arrival we discovered they had been dispatched elsewhere to support the newly thriving tourism business. We had been promised the help of local divers for support, but Buddy Quattlebaum's dive shop was simply too busy to free up any staff to help us. I was a little perturbed—we knew hundreds of divers who would have eagerly joined us from the U.S. But we soon resolved that we could do our dives without a big team. Paul and I had plenty of equipment, all our expedition supplies, and a solid emergency recompression plan in case something went wrong. Buddy helped us get established in the jungle and assigned a shop staff member, Gabriel, as the standby contact person for emergency support. We could reach either of them by radio if we were desperate.

Our logbooks had grown fat with many thousands of dives, and I was feeling a mix of enthusiasm and invincibility. I had recently been inducted into the inaugural class of the Women Divers Hall of Fame, and my career was riding high. Consulting work, photojournalism assignments, and television opportunities were on the rise. Paul and I still struggled with finding our relationship amidst a hectic work and diving life, but we were trying. I tried to make a cozy, traditional home life for us—I kept a clean house, made gourmet dinners, and otherwise attempted to create normalcy—but I felt a bit like an impostor in those efforts. I was much more comfortable cooking on an open fire or hiking in the bush.

On this return to Mexico, Paul and I were excited about having time together without the noise and distraction of work, but the trip was mostly about chasing the lead we had left at the bottom of the Pit. We both knew it was only a matter of time before somebody else discovered what we had uncovered there.

We fixed an aluminum ladder to a tree that was precariously hanging from the lip of the cenote and climbed down the twenty feet to the water's surface. The water was remarkably clear, and as soon as my face was immersed, I could see the huge subterranean cavern shaped like a stadium, with multihued limestone running in horizontal bands around the periphery. About a hundred feet below us was a smoky layer of hydrogen sulfide, with plumes of white sulfurous fumes rising through the water. It looked like a simmering witch's cauldron. As Paul and I descended, we found ourselves passing through shimmering haloclines, where freshwater meets salt. My eyes struggled to focus through the haze. Then we reached the sulfide layer and descended through fifteen feet of milky water that smelled like rotten eggs. Below that layer, the gigantic cavern was fully revealed in a zone of crystal-clear, ice-blue salt water.

At 180 feet, we reached the base of an enormous conical mound, a tangled mass of boulders, silt, trees, and debris that had fallen through the skylight opening above. At the base of the pile, we noted the spot where the first explorers of the area, Kay Walten and Dan Lins, had ended their journey five years before. I recognized the spot from Kay's descriptions and noted a small triangular marker she had placed on the termination of her guide line. From there, we followed the line we had left behind three years earlier. It was slightly stained but bore no evidence that anybody else had found it.

The helium mixture eased into my rebreather with a squealing hiss. I reached forward to trigger the motor on my scooter and picked up speed in an enormous descending room I had named the Cardea Passage, after the Roman goddess of hinges. I felt I was approaching a turning point in my life, descending into the darkness of an unknown future. The last daylight dwindled behind us as we quickly motored to the point where we had ended our previous exploration. It was incredible to take in the vastness of the space around me. On our earlier exploration, we had been diving with an air mix and were high on nitrogen narcosis. This time, with the clarity offered by helium, I noted every detail of the vaulted room and the colossal limestone boulders on the floor.

At 310 feet of depth, Paul tied into the end of our previous exploration line and pushed through a silty restriction. Breaking into a new tunnel and laying line for the first time is the ultimate thrill in underwater cave exploration. It's exhilarating, an experience like no other, and for many divers it is the reason we willingly expose ourselves to so much risk. Now, Paul and I entered an even more enormous room, larger than any we'd ever seen before. It was just a black void, without any defined edges. Our lights illuminated only a shadowy sphere, and the blackness extended as far as we could see in all directions. I was elated, my heart thrumming with excitement. There was so much more to discover! We were in an immaculate place no human being had ever been before. I heard Paul go "Whoop!" through his rebreather, and a thrill tingled through my body, bringing every neuron to full attention. For the first time in quite a while, I felt like I was in perfect synchronicity with Paul.

While Paul spooled out the fresh line, I carefully recorded distance measurements, counting the knots we had tied into our guide line at ten-foot intervals. Every few minutes I paused to take a new compass bearing and jot down a description of some of the formations around me—an odd-shaped boulder, an eroded cubbyhole, or perpendicular passages that we might want to investigate later. As Paul pressed ahead, I hurried to fill in the details of our survey while keeping him in sight. From our earlier sketches, we knew we were not far from the Blue Abyss, a spot in the adjoining cave system where the floor suddenly dropped down to 260 feet. If we found the end of the exploration in the Blue Abyss, we'd know we had broken through and connected the two longest caves in the world.

Gradually the walls of the enormous room tapered into another narrow restriction. We eased off the triggers of the scooters and looked at the pinch point to see if the cave was ending. Even though the gap appeared small, it is never claustrophobia that holds a cave diver back. We are only trying to figure out if we can squeeze through a space and return without getting stuck. Paul went first, forcing his body and gear through the tiny slot. I waited for a moment to see if he would turn back. With no sign of his retreat, I scrambled through a curtain of silt and crumbling rock to emerge in yet another giant space that spread before us. Spooling out the line, we pressed on until we reached the back of that passage, where the room ended in another tight constriction. Now we were at 390 feet, deeper than any known spot in the Yucatán. Whether it was intuition or optimism, we felt confident that the connection had to be nearby.

But we had been down about an hour, and our time was up. Finding the hidden link would have to wait for our next dive. Paul cut the line with his Z-knife and carefully tied off the end to a rock. Divers call the final tie-off the "bitter end" for good reason: it's always a frustrating moment. There was still plenty of cave calling us onward, but we were looking at about four hours of decompression on the way back up. That day's adventure was finished. Perhaps after a rest day, we could try again.

We retraced our journey, methodically working our way back through the chain of rooms and restrictions, precisely the way we'd come down. We started our decompression stops at 260 feet and planned on stopping every ten feet all the way to the surface. At each level, we hung patiently beside the wall of the cave, allowing our bodies enough time to offload the helium and nitrogen.

Reaching 60 feet from the surface, I saw a pillar of sunlight streaming down through the water like a laser beam. I switched off my diving light and enjoyed the spectacle. Everything in that moment was tranquil and beautiful.

That's when I felt it.

I had always prided myself on paying attention to the smallest physical changes in my body. I liked to think of myself as someone who was alert to potential danger, and when you're diving as deep as we did, there's always a danger. At first, it was just an odd sensation I felt, an awareness that something was not quite right. I could feel a tingling on the inside of my thighs, like rows of tiny ants on the march. Could it be actual bugs? Had tiny chiggers found their way into my suit unnoticed? In the past, I had unintentionally entertained scorpions and cockroaches inside exposure suits, but this felt very different. Could it be caused by contact with the poisonwood tree? No, the weird itchy feeling on both my thighs felt far too symmetrical. Denial began creeping into my head. I tried to will the sensation away, but with every minute that passed it felt stronger.

I reached down to touch my legs. My dry suit felt a little tight, but otherwise my legs seemed normal. For a moment I was reassured. And then the tingling came back, worse. The ants had multiplied into an army swarming over the area from my knees to my hips, burrowing into the layer between my skin and muscle. I could imagine them biting and devouring my tissues, and now I knew there was only one explanation: I was bent!

It's hard to exaggerate the terror of getting the bends. Every diver knows about it, we've all read or heard about it, but it seems remote and unreal. I knew that it often began with a minor symptom, then suddenly erupted into a full-blown crisis. One moment you're having trouble feeling your fingertips, then suddenly you can't move your arm, and then you can't walk. In one afternoon, the bends can end a career. Or even a life.

I had never felt a sensation like this before. Despite having taught hundreds of divers about the symptoms of decompression illness, nothing prepared me for the absolute horror of recognizing that it was happening to me. I could visualize the tiny bubbles of helium and nitrogen wreaking havoc in my tissues. Under pressure, my tissues had soaked up the gas like a sponge soaks up water, but now the tiny bubbles were expanding in my body as I ascended. If I didn't get the situation under control, I was not just a ticking time bomb; I was a shaken pop bottle waiting to see if the cap was going to get ripped off.

I quickly decided to lengthen my stay underwater. The extra time under pressure might give the rogue bubbles the time they needed to get reabsorbed and breathed away, but the extra hang time left me feeling very chilled. Paul was swimming laps around the cavern to stay warm, and I was sitting still, analyzing every sensation that confirmed my self-diagnosis. Moving might have helped me stay warm and might have ramped up my body's efforts to deal with the invading bubbles, but I felt a crippling terror, afraid that if I moved, I might set a bubble loose in my circulation that would paralyze or kill me. That's what makes the bends so insidious. A diver might only ever experience a minor rash, but someone else could be unlucky enough to have a single bubble get trapped in a very bad place like the spinal cord or brain. Early symptoms are minor but when they start getting worse, you are left with a sense of impending disaster.

My mind was racing. I kept going over the details of my dive, trying to figure out what had gone wrong. Most people don't even realize they are bent until after they've reached the surface or are completely out of the water. The symptoms tend to creep up during the first hour, and sometimes only after several hours. Some people get all the way home after a great day of diving and don't recognize the growing issue until they are unable to fall asleep from the pain and discomfort. If I was having symptoms while still underwater, I must have it bad.

Within the cave-diving community, getting bent inevitably brings shame upon the diver. To a lot of divers, getting bent is something that only happens to amateurs and the incompetent. When people find out that someone has gotten the bends, they immediately start picking apart that person's dive, making judgments about the choices they'd made. They criticize a dive plan, or gas or equipment choices. They debate the chosen mathematics used to calculate a decompression schedule. They even find fault with particular brands of gear. I've seen instances when someone has died in a diving accident, and people on the internet were dissecting their supposed mistakes even before the family was notified. In that moment, suspended in the sinkhole, with my legs on fire, I could already feel the sting of criticism. One moment I was a heroic explorer, pushing the limits of physiology and technology, filled with self-importance. The next moment I might lose the respect of my community. I had worked hard to get to this point in life and feared this random attack of bubbles would wipe away everything. Diving suffused my life, made my career viable, and fueled my marriage. Without diving, nothing would be left.

I wanted to tear off my dry suit and scratch my thighs. I thought that removing the suit might help me feel better. Perhaps I could see evidence of what I was feeling, like bruising, redness, or a rash. At least seeing the signs might make me feel in control of the situation. Buddy Quattlebaum had installed a large rain barrel in the cavern. The big barrel was inverted on the flat cave ceiling at twenty feet of depth, and it was large enough to accommodate a seated diver on a wooden plank with their legs dangling in the water. The makeshift air bell was a place to get dry and warm and was stocked with a supply of food and water in case a decompression stay was lengthy.

While Paul continued swimming laps around the cavern, I swam to the bell and worked through a plan. I couldn't even think about communicating my problem to Paul. I wanted the pain to go away, and I could focus on nothing other than myself. I took off my emergency bailout tanks and clipped them on hanging string loops. I removed my rebreather and inflated the buoyancy wing so it would float inside the capsule. I pulled my cumbersome body up onto the plank inside the barrel. I knew I had scarce time to figure things out. Every exhaled breath would raise the carbon dioxide level in the confined space, and eventually it would be high enough that I would risk losing consciousness. With great difficulty, I heaved the rebreather onto the plank beside me and tried to twist my torso in a way that would enable me to breathe from the unit while sitting on the plank. It was precarious. If I lost balance, I would tumble into the water and have to start the process all over again. I tried to ease my way out of the top of my dry suit, but the physical effort was exhausting, and my body heat was raising the temperature in the barrel to an uncomfortable level. Light exercise is beneficial for decompression, but I was overexerting myself trying to organize my gear and breathe efficiently from it. I drank the jug of water, redressed myself, and dropped back into the water. I knew that I needed Paul's help, but I didn't have the energy to gain his attention. My every thought focused on my own body and the strange sensations I was feeling. I was terrified.

The niggles in my thighs weren't going away, and I was beginning to feel unwell. I knew what was happening to me but I was less and less in a position to take care of myself. Even though Paul was only a few feet away, he had no idea I was in trouble. At times, he swam so close I could have reached out and touched him, but something held me back. The chasm between us at that point is hard to explain. I think I wanted him to notice that something was wrong with me and rush to my side, but he seemed incapable of seeing that I was ill. Perhaps he was in denial too.

I filled my dry suit like a Goodyear blimp and pinned myself, motionless on the ceiling at twenty feet, and I lengthened my decompression stop again, hoping the extended time under pressure would relieve my symptoms. I struggled to ignore the primitive instincts that were screaming at me to get out of the water, to get the dive over with.

At the end of an extended decompression hang and extremely slow ascent, I reached the bottom of the aluminum ladder. It had been more than five hours since I had first dropped below the hydrogen sulfide and swum into the cave. The relief I felt was laced with uncertainty. Paul had already climbed the ladder with his gear, but I removed my equipment one piece at a time and left everything floating on the surface. This was not the time to make a heroic exit carrying everything up the ladder. As I gingerly pulled myself up the ladder, a profound and intensifying exhaustion told me I still had a serious problem. I had never been so tired after a dive. I felt the heavy dread rippling through my leg muscles with every step up the ladder. I was trying to push back the rushing panic and concentrate on getting to the top without falling.

I could barely lift myself up the final rungs. At the top of the ladder, I rolled sideways and collapsed onto the dirt. I was unable to get out of my dry suit. My resolve eroded. I lost the ability to help myself and asked Paul for help getting the zipper open and the suit peeled away. I collapsed on my sleeping mat and hoped everything would go away. I was still terrified. This could be the end of my career, or even the end of my life if things continued to get worse.

In order to flush away inert gas and improve my situation, I breathed a rich oxygen mix from a scuba regulator, and I drank as much water as I could handle to combat the associated dehydration. Then I closed my eyes and silently suffered. My body was compromised, but my bladder was working fine. That meant I frequently had to roll off my sleeping mat and squat in the jungle. It took all the energy I could muster, and each time I had to urinate I had to stop breathing the rich, life-saving oxygen mix that was flushing away my bubbles. When I returned to the mat and started back on the gas, I noticed that a weird mottled spiderweb of bruises was erupting on my thighs and biceps. My belly itched and burned too. While I was breathing from the tank, the symptoms improved, but as soon as I stopped, the bruising grew worse. The area spread and darkened into a purple blotchy mess. "My god, what have I done?" I wondered. It looked like I was bruising from the inside out, and I felt like I had been beaten with a baseball bat. I was still too exhausted to communicate with Paul, who was sitting only five yards away at the fire. I wished he would sweep me up and make it all go away. Was our bond so weak that he could not even ask me what was wrong? I wanted to sleep but I could only close my eyes and surrender to the pain. I could not have fought my way out of a wet paper bag. I was giving up.

After emptying the scuba tank, I lay in silence under my mosquito net and listened to the chirps and howls of the jungle. I had breathed the oxygen mix for over an hour. I felt a little better and almost thought I could sleep, when an eerie visitor scared me back to reality. A stout black scorpion, the size of the palm of my hand, was dangling from the interior of the netting, working its way toward my face, brandishing its lethal tail. Its arrival was enough to help me find the strength to get up and walk to the campfire where Paul sat. The late-afternoon sun was illuminating horizontal bands of smoke through the trees.

"I'm bent" was all I said.

"No shit, Sherlock," he responded. "What do you want me to do?"

Though I doubt he meant his words to be hurtful, I was stunned. I needed someone to take charge. I needed comfort, reassurance, and guidance. I wanted my indomitable French-Canadian husband to sweep me into his arms and make everything better. I wanted him to recognize the severity of the situation that could end my diving career. Perhaps, like me, he realized that diving was the only glue that held us together and he was already moving on.

He seemed more intent to sit alone by the fire than to comfort me at the most frightening time in my life.

Those three simple words—"No shit, Sherlock"—revealed all I needed to know about our relationship. Paul was my dive buddy, and not what I needed in a life partner. I wanted to curl up into a fetal position and cry, but given his seeming indifference, I had to take charge.

"Take pictures," I responded coldly. "If this is the reality of expedition life, then we should document it. And call Gabriel on the radio and tell him what's happening. I'm going to need to go to a recompression chamber, but I can't even consider walking now. I'm going to treat myself in the morning in the water."

When evacuation is not quick or even possible—when you're far back in the jungle, you can't exactly call an ambulance—some divers try emergency in-water treatment, compressing themselves back underwater while breathing oxygen-rich gas. Rebreathers were perfect for doing that. Generally, divers would not risk breathing such a high concentration of oxygen, but given the circumstances, the potential gains for healing my damaged tissues far outweighed the risks of having an oxygen seizure. It is a risky move without support, but it was the only way I could regain enough strength to get myself out of the jungle. The earlier this strategy can be deployed, the better the outcome for the diver, but I just couldn't manage making the effort in this mentally debilitating moment.

Paul passed me the radio handset so I could speak with Gabriel and Buddy. "Hey, I'm going to try to recompress in the water as soon as I can get in," I told them. "When the sun comes up, send someone else to take my place, to help Paul with the rest of his dives. I don't want to be the one to wreck Paul's expedition." I was already feeling humiliated, convinced all this was somehow my fault. I should have asked Paul for help, but I was more worried that he'd be mad at me about ruining the expedition. It was clear I was seriously hurt, but he did not seem to be able to help. And so I organized a replacement dive buddy for him, and he seemed fine with that. All I could think was that the dive expedition must be more important to him than me.

I dozed on and off through the night, and as the jungle canopy came alive with the morning birds, I was feeling a little more energetic. I still had some soreness and inflammation in my thighs and biceps. My bulging, tender arms looked like Popeye's—though without his characteristic muscle tone and anchor tattoos. My whole body was swollen and puffy, and I looked as though I had gained twenty pounds. My moon face was droopy and sad, and my hair was matted with smoky soot from the fire and greasy insect repellent. Wearing a long, dirty T-shirt and flip-flops, I plopped my body in the camp chair by the smoldering fire.

We didn't talk much in the morning beyond the practical plan for what would come next. I ate a bowl of instant oatmeal and Paul helped by prepping my rebreather. I slowly dressed in my musty dry suit and wobblily climbed down the aluminum ladder to get back in the water. I would spend two hours guzzling oxygen at depths down to forty-five feet. Even if it was too late to recompress bubbles, this treatment under pressure would send a healthy dose of concentrated oxygen to my damaged tissues—more concentration than I could get from breathing from an oxygen tank on the surface.

Not wanting to consume his dwindling expedition gas, Paul snorkeled overhead, on the surface, standing by in case I had any adverse effects from the treatment. I wished my husband had been there to reassure me, face-to-face underwater. I was truly terrified and felt completely alone, and looking up at his silhouette on the surface offered no comfort. I kept thinking about Jason Richards and his oxygen seizure at Wakulla Springs. Had it not been for the quick action of his diving partners, he would have drowned. I remembered seeing his blue toes hanging off the end of the gurney as he was loaded into the ambulance. I did not want to go that way.

Two hours later, my makeshift treatment was complete. I felt a vast improvement in energy but still wanted to wait for an escort to walk me out of the jungle and back to the road. While I waited, Paul decided to go off on a solo exploration dive. I sat in my camp chair alone as the sky became progressively darker, as though desperation was closing in. And then the sky literally opened up. A torrential storm dropped buckets of rain while I sought shelter under the kitchen tarp. The water poured off the edges of the sagging fabric and the canopy gave in to the rain and collapsed, taking with it every bit of courage I had left.

The rain slowed to a drizzle, and Gabriel and Buddy finally arrived in an old Jeep, pushing it deep into the woods on the bumpy trail. They offered me a ride out, but everything on my body hurt and I knew that jostling in the Jeep would be more painful than a slow amble along the rocky trail. I struck out walking gingerly along the path, Gabriel beside me, while Buddy drove ahead. We soon discovered that concerned friends had left us a jungle truck at the halfway mark on our way out, where the rocky trail led to a much better dirt road. I eagerly got in, and Gabriel drove the smoke-belching, open-cockpit beast while I sat bravely beside him getting soaked from head to toe in mud that sprayed up off the giant tires. The day before we'd experienced remarkable discovery; now all I felt was abandonment, humiliation, and defeat. Gabriel stepped in as my guardian angel, and although he was kind and thoughtful, he was not my husband.

Gabriel drove me straight to the hyperbaric facility in Playa del Carmen, where I met with Dr. Andrés Medina. I thought my ordeal would soon be over, but there were already two patients locked in the chamber in the middle of a treatment that would span six hours. Dr. Medina was shocked to hear about the depths and length of my dive and even more surprised that we had conducted in-water recompression.

"Where on earth did you come from?" he asked.

Dr. Medina was accustomed to treating casual holiday divers and had no background in technical diving. I pulled out my note-book and shared the remarks from my dive. I had written out a detailed spreadsheet of depths, times, and gas choices to approx-imate the data that was stored in my wrist computer. I explained how our rebreathers worked and how they had helped me manage my emergency treatment. After he had listened to the whole story, he offered to send me to hyperbaric facilities in distant Cozumel or Cancún for more immediate treatment, but my journey had already been too long and painful and I couldn't imagine traveling even farther. Through tears, I told him I would wait for his cham-ber to become available.

Later that evening the recompression chamber was finally ready for me to enter. It looked like a mini submarine. The large steel tube had a thick round metal door on one end that sealed with a heavy, spinning latch. I changed into medical scrubs and grabbed two large bottles of Gatorade for the lengthy treatment. I crawled into the hatch and crouched to pass through the second locking door. The main compartment was not much longer than my body. I reclined on the cold vinyl mattress on the folding bunk while the hyperbaric nurse, Elmer, fitted me with an aviator-style mask. I stretched out and tried to get comfortable while he checked my blood pressure and pulse, which was bounding from nervous-ness. After passing me a set of protective headphones, Elmer slammed the heavy steel door shut.

Despite having helped run a chamber, and spending long hours in one at Wakulla, lying helplessly on a cot as a patient was a completely different experience. My first of three chamber treat-ments would require five hours under pressure, breathing pure oxygen up to a pressure equivalent to being sixty feet underwater. That concentration is more than double the recommended maxi-mum oxygen exposure for diving, and that meant an assistant would need to be with me in case anything went wrong. Seizures, burst eardrums, and fire are all risks in recompression chambers, beyond the medical issues I was already facing.

"We're going to pressurize you now, so be ready to equalize your ears," Elmer said as the hiss of gas filled the space around me.

I pinched my nose and wiggled my jaw to equalize the pressure in my middle ear, knowing that if I failed, Elmer would have to lance my eardrums to prevent a more dangerous situation. The chamber got very hot, and I wasn't sure if my sweat was a nervous flush or from the heat of compression. As the chamber pressure increased, Dr. Medina announced the depth through a crackling speaker: "Twenty feet . . . thirty feet . . ."

The noise was deafening, and Elmer pressed his palms over his headphones. Recompression therapy is perhaps the only medical treatment where your technician or nurse is subjected to the identical treatment that you receive. For me, the increasing pressure was squeezing down any remaining gas bubbles in my body. For Elmer, it was the equivalent of going for a dry dive. We both tried to equalize our ears quickly and I noticed my heartbeat through my sinuses and eardrums.

"Forty feet . . . fifty feet. Just ten more."

Then the noise stopped and Elmer pulled off his headset. "Are you okay?" he asked.

I assessed every aching part of my body. "I think so."

I felt a wave of relief. My symptoms had improved: the pain had lessened, the swelling had reduced, and I had more energy and optimism. I was beginning to feel as though this whole ordeal was over. At three in the morning I walked out of the clinic alone and into the humid darkness of Playa del Carmen. I lay down in my camper van, which Gabriel had parked across from the clinic, and I crashed hard for two hours, waking up only when the pain in my body was worse than before my treatment. My biceps were swollen again. My stomach muscles were sore. My thighs were burning. My hips were hurting. My shoulders were clicking. I rolled to one side and then the other, but could not find a comfortable spot.

As the pain intensified, my spirits plunged. I was exhausted but couldn't sleep—my blanket and sleeping bag were in the jungle, and all I had was a small jacket to wrap around me, and it was chilly. Finally, too cold and uncomfortable to lay still, I got up and roamed the predawn streets. Stray dogs dug through garbage in the alleys and shopkeepers swept dust into the streets. With no traffic lights to help, delivery trucks honked their way through chaotic intersections guarded by policemen with whistles. I wandered for hours, shuffling aimless feet on a purposeless route.

At first light, I drove to Buddy's dive shop and used his radio to call Paul in the jungle camp. The clinic would not open for hours, and I needed emotional support. All the numbness had morphed into pain, and I was an emotional wreck. Paul agreed to come out of the jungle and meet me at the chamber.

An hour or so later, a weary Dr. Medina turned up to meet me at the clinic door. "Come on in!" he said. "How are you feeling?"

"Terrible! I feel worse than when I came in yesterday."

"Fantastic!" he replied.

"What? That's cruel! My career is over!" I said, choking back the tears.

He gently placed a hand on my shoulder. "We'll talk about that later, but right now, pain is a good sign," he said. "Pain means you are healing. You are reawakening the damaged nerves and vessels of your body. This is progress. Yesterday's neurological deficit is improving and the pain you are feeling means healing." Before my time in the chamber the day before, Dr. Medina had detected a significant inability to sense touch. He had poked me with needles and stroked my skin with brushes. He had checked my reflexes and looked in my eyes. I thought the swelling was deadening my sense of touch, but Dr. Medina was convinced I had suffered a serious form of decompression illness, one that affects the spinal cord.

I was shocked. "Neurological decompression sickness? A spinal cord hit?" This didn't sound like something with a happy ending. Decompression illness can be treated, but it is hard to predict the final outcome for most patients. Rogue bubbles that lodge in the spinal cord can cause terrible damage. But so far I was lucky in some ways. Some people never walk, let alone dive again.

I sat on the examining table while Dr. Medina went about preparing the chamber for another run. Knowing Paul would be there when I finished this six-hour session helped me hold things together, but just barely. When I actually saw Paul's face in the porthole five hours into treatment, I felt that I could finally share the burden. He smiled and waved through the tiny window, which eased some of my tension for my final hour in the chamber. His slight smile and look of worry let me know that he was now on watch. I could stand down and rest for a while.

By eleven that night Paul and I were asleep together in a hotel room. It was good to take a hot shower and properly rest in a real bed. But I awoke a couple of hours later once again feeling like I had been run over by a truck. I was scared, of course, but since "pain is good," I tried to take it as a blessing. After breakfast, we visited with the doctor again, and he decided that I should have one more, shorter treatment. My symptoms had continued to improve, so Paul hiked back into the jungle for more diving while I crawled back into the cold steel tube. Afterward, I spent the afternoon wandering the streets of Playa wondering what would come next for me in my career and in my marriage. I found a natural food store and knocked back a fresh carrot, beet, and spinach juice while I waited for the customary intensification of pain. This time, it never materialized. I sensed a small twinge of confidence that I was getting better.

I chose to splurge on another night in the hotel despite offers from Gabriel and his girlfriend to stay with them. Not surprisingly, I just wasn't feeling social. The owner of the hotel was playing beautiful classical guitar, and I could no longer resist the exhaustion of the past five days, and I finally had a restful sleep.

My morning of reckoning came with Dr. Medina. Although I still had some pain, he declared that my progress was good and that further treatments might compromise my lungs. Adding all the treatment oxygen on top of diving had made breathing uncomfortable. My lungs agreed that it was time to go home.

And that's when he said it: "Never dive again."

Three powerful words, like the ones Paul had spoken in the jungle. They morphed into an earworm: *No shit, Sherlock. Never dive again.* Who was I? What was I? A technical exploration diver and a loving wife; each undone by three simple words.

The doctor must have seen the panic rising in my face, because he quickly began talking. "I have never treated anyone from such a deep dive. So I don't know what to tell you. Statistically, you are more likely to get bent again, and any sane doctor would tell you to just avoid diving. But I will leave it up to you."

Hyperbaric physicians rarely get a chance to treat technical divers who have been using exotic gas mixes like helium at extreme depths. Even fewer physicians will ever treat a woman diving in this range. My five-hour dive profile was an anomaly in Dr. Medina's experience: I was a technical diver, using a rebreather, mixing weird life-support gases, and I was a woman. There were no medical journals that offered advice to deep-diving women or the physicians who had to treat them. I was on my own. Dr. Medina had helped me heal, and he would continue treating vacationers and instructors in the holiday paradise of the Riviera Maya, but he might never meet another patient like me.

I made a very slow walk back to the jungle camp to give Paul the news. He'd probably still be in the water, but I could at least be useful packing up camp and feeding him when his dive was over. The walk took a couple of hours and I had to pace myself, feeling the exertion of the hike and the heat. My aches and pains intensified with hard work, but I appreciated the freedom of the jungle. The sounds of the motmot birds and parrots cheered me up. The smell of wet leaves and smoldering campfires reminded me how much I loved this place. I was determined to walk back to health and move cautiously in the direction of everything I loved to do. I could not imagine giving up diving without a fight.

Paul took the news in stride, but I couldn't get a read on his emotions. I wasn't sure whether he was frightened about losing a dive partner or happy to get a wife who would slow down and stay home. His emotional detachment left me feeling lonely, and I wished he had been better able to talk about feelings, fears, and the future. I was afraid to ask what he was thinking for fear I might not like the answer, so instead I told him I would take life one day at a time for a while. I still had plenty of recovery time ahead before I could get back in the water.

It was almost two weeks after the incident when we finally arrived back in Florida in our van. Friends, store customers, and people I didn't even know had been calling the dive shop for news. I was determined to be as upfront about my injury as I could, given that some divers had already cruelly weighed in with misinformation online. Radio transmissions from the bush had been overheard and relayed up and down the Yucatán coast even before I had made my effort at in-water recompression. "I'm too tired to walk out of the jungle" became "Jill is paralyzed in the jungle." The inaccuracies magnified with every retelling. But I wasn't going to let the feelings of shame or embarrassment hold me back from going online to share my own version of what happened.

While some people were gossiping about me, I was also overwhelmed with supportive messages from other divers in the community who had experienced what I'd gone through. I learned that getting bent was a lot more common than I had imagined. Some people I'd known for years only now told me it had happened to them. One close diving friend had even been to the recompression chamber seventeen times. But most offered their reflections in a hushed whisper. It was their well-kept secret.

I DIDN'T FOLLOW the doctor's advice. Two and a half months after the accident, I squeezed into a wetsuit and got back in the water. But something was different. I had both lost and gained on that Mexican trip. What I'd lost was a feeling of invincibility, though at the time I would never have put it that way. With thousands of dives under my belt, I'd never really thought that I would get bent. Nobody could tell me why I got bent—body fat, dehydration, my cycle, tight gear, or simply the luck of the draw could have all contributed, but there was never a definitive conclusion. Any feelings of invincibility had dissipated, not just because I'm statistically at a greater risk of getting bent again but also because I know that when it comes to exercising caution while diving, there are no shortcuts. All exploration involves danger and risk, and as cave divers, we certainly accept that. But the lesson is clear: we must know our limits.

As for the missing link, it seems irrelevant now. In 2018, the longest underwater cave first dwarfed and then absorbed the Pit, the Blue Abyss, and other systems, creating a mammoth super-cave more than 215 miles long. As exploration continues, it is becoming clear that all caves of the Yucatán Peninsula are connected. Perhaps my link to the Blue Abyss lies in one of the dark cubbyholes or side passageways I found with Paul in the Pit. Maybe it's somewhere in the sub-basement of the Blue Abyss, hidden among the rocks and boulders. In the years since Paul and I were in the Pit, a vibrant cave-diving exploration community has blossomed in Mexico. I hope one of those divers will discover the deep link. The exploration and the connection rightfully belongs to them.

ICE ISLAND

2001

IN EARLY 2001, I was still coming to terms with my personal limits, and that's when global climate change nearly killed me.

A gigantic chunk of ice more than 180 miles long broke away from the Ross Ice Shelf, near McMurdo Sound in Antarctica, instantly creating the world's largest iceberg. The rupture of B-15—or Godzilla, as the scientific community nicknamed it—touched off an intense debate about whether rising temperatures could eventually cause the polar ice caps to melt. This drifting behemoth brought global warming to the top of the news cycle.

On a steamy May long weekend the year before, I had found myself at yet another annual cave-diving conference. It was not long after returning from the Pit and it was nice to be back in the community. Ignoring Dr. Medina's advice, I was easing my way back into diving too. I stepped away from the podium, my presentation on my world-record dive at Wakulla Springs complete, and made my way to the Holiday Inn's pool deck to join the line for the buffet lunch. A familiar voice called, "Jill!" It was Wes Skiles. He squeezed into the line ahead of the hundred or so attendees waiting for food, but nobody seemed to mind. We piled warm potato salad and cold sliced turkey onto flimsy paper plates and sought shade under a thatched umbrella.

Wes was wearing his standard Florida uniform: worn-out

flip-flops, a sleeveless surfer's T-shirt, and an old pair of comfortable torn jeans. His auburn mustache and beard were tinted with streaks of gray, but when he smiled, his eyes twinkled with youth. My backwoods brother's voice bore a lazy Southern drawl that was accentuated by his nasty habit of suckling on a jaw full of oily chewing tobacco. We shared a love of cave diving and adventure and were nurturing our new creative partnership. He had recently hired me for several reality television gigs, and I had been working hard to shift into full-time TV production work with him. We'd been filming alligators in swimming pools, sharks attacking swimmers, and pseudo-heroic stories about the "scariest encounter" or the "greatest survivor" to quench the new American thirst for adventure television. We'd been brewing up ideas for something better and more meaningful, something beyond the reward of a videographer's paycheck for diving.

"National Geographic called back about Antarctica," he said. "They're interested in hearing our ideas about diving."

"You've got my attention!" I said with a big grin.

"I need your help," he said. "I need you to teach me how to ice dive. I need you to lead the dive team." I was surprised by his announcement. Until that point, I had been his young apprentice. I was also nervous about my slow recovery. Although my dives were gradually increasing, I was worried about getting bent again. But knowing that Wes had been bent and recovered on several occasions made me more comfortable about what lay ahead for both of us.

Wes was born in Jacksonville, Florida. His mother, Marjorie, fostered a love of outdoor adventure in her boys and frequently pulled Wes and his brother Jimbo out of school when the ocean surf was peaking. She would drive up to the door of the school with surfboards on the roof rack, then tell the principal she needed to have the boys excused from class. They were expected to work hard in school, but Marjorie wanted them to have a life balanced with real adventure, and that's what Wes had. But although he was a world-class expedition diver and filmmaker, he had no experience in cold water. For me, it was my element. Even a casual dive in Canada is bone chilling, and I actually enjoyed the challenge and beauty of ice diving. Would I lead the dive team to shoot a National Geographic project in Antarctica and teach Wes to ice dive? It was a big deal for me, and I would persuade Paul to join us too. Wes and I had put a lot of research time into the initial pitch. If National Geographic put up the funds, I would also have the opportunity to write my first documentary film.

I was overflowing with excitement when I broached the expedition with Paul that evening. "Honey," I began, "Wes wants us to work on a big project with him. He's also asked me to write and produce the film."

"That sounds like *real* work," Paul said enthusiastically. "What's the project?"

Knowing that Paul, although born and raised in chilly Quebec, had no love for cold water, I teased, "Well, it's in the south . . ."

"Where is it, Jill?"

"We can get away from our dive shop in the slow season . . ."

"Come on, Jill, where is it?"

"It's Nat Geo and properly paid."

"What are you not telling me?" he demanded.

I relented. "It's the Southern Ocean."

"Where is the Southern Ocean?" he asked.

"Antarctica . . ."

"I'm not too excited," he admitted. "This is going to kill me."

Paul's coldest dive so far had been a brief visit to the St. Lawrence Seaway to dive the wreck of the *Empress of Ireland*. In 1914, more than a thousand people had perished when the liner went down in fourteen minutes. He loved diving the wreck but was not very happy about the discomfort of the cold water. Yet the St. Lawrence was almost twenty-five degrees warmer than the waters off Antarctica, and the topside conditions near the South Pole would be far worse than anything he had felt before. To make matters even more challenging for a man who suffered from seasickness, the entire project would be boat-based. But he, like I, thought this would be an incredible adventure as well as a decent paycheck. Perhaps our marriage was now transforming to more of a business partnership? The romance had waned, but the partnership was convenient. The intensity of love was missing, but we could still work together and enjoy the experience.

My perspective on the adventure ahead had additional undertones. For some, a brush with death or fighting off a ruthless attacker might forever destroy their confidence. For me, there were weird gifts attached to my life's most terrifying experiences. Fighting off a burglar in my home helped me emerge from my younger self. Enduring online critics when I was struck down with the bends helped me to accept rejection. I had slowly regained my mettle one dive at a time after the incident at the Pit. The Pit project helped me build a more mature approach to planning and safety. Even Henry Kendall's death offered me important lessons. At seventy-three, he was living life at full speed. I know he certainly didn't want to die that day, but he wasn't afraid to live fully on every day he was gifted. I wanted to live like Henry.

I looked on failure as something I called "discovery learning." Rejection from outsiders became affirmation that I was doing something unusual and even remarkable. I knew that boundaries placed in front of myself and other women only shielded the status quo. I looked at gatekeepers who stood in my way as people who lacked vision. I was determined to explore and experience the unimaginable because I was convinced that I was capable of more than I could imagine.

When you are a dreamer, it is a good idea to surround yourself with other creative souls. Dreamers are happy to contribute wild ideas and collaborate with others. They are prepared to win and lose arguments and work together to morph an idea into something tangible. Dreamers see potential. Dreamers learn to jump into cold water, metamorphose, and do things that are uncomfortable—things that have never been done before. Wes Skiles was my partner in dreaming. We catalyzed each other's ideas into creative success. Together, we could brainstorm, refine, and construct a plan to do remarkable things. Where one of us lacked skills or ideas, the other filled the void. And he was helping me to see that I had to let go of societal expectations and that I had to stop focusing on changing myself to make Paul happy. If I wasn't happy, I was wasting my life.

When Wes and I prepared our pitch to National Geographic for our Antarctica project, we had only a skeleton of an idea. "Following the historic path of English polar explorer Sir Ernest Shackleton's expedition across the Southern Ocean" was hardly a complete narrative for a film. We read news stories, spoke to scientists, and delved into polar history. I got familiar with the geography of the continent. I looked at satellite photos of glaciers and ice sheets and learned about the aquatic life we might encounter.

Both Wes and I admired Shackleton as one of the greatest leaders of all time. We swapped his books, *South* and *The Heart of the Antarctic*, and respected this man who faced outrageous odds to rescue his crew after his ship was crushed in the ice pack. During his Imperial Trans-Antarctic Expedition of 1914–17, his vessel *Endurance* became engulfed and was slowly crushed before sinking and leaving the crew stranded. First camping on the sea ice, the men eventually launched lifeboats when the pack broke up the following spring. Then they made an improbable 720-mile journey across the vicious ocean to land on Elephant Island. Stranded again on an uninhabited rocky shoreline, Shackleton built a camp out of overturned lifeboats and sailed bravely onward, finally landing on South Georgia. He scaled treacherous ice fields and a difficult mountain pass to reach help in a whaling community, and was then able to rescue his men after almost two years without shelter or support in the most hostile climate on earth.

This epic story of exploration and leadership had also marked the passing of an era. One could argue that, as a member of the post-boomer generation, I missed the age of exploration. The earth was "conquered." The map was complete. The poles had been discovered. Sir Edmund Hillary and Tenzing Norgay had climbed Everest. Don Walsh, Jacques Piccard, and even filmmaker James Cameron had touched the blackness of the ocean's deepest trench, and astronauts Neil Armstrong and Buzz Aldrin had walked on the moon. What did we have on Shackleton? Would his epic feat overshadow anything we could do there?

I didn't think so. As I saw it, the age of exploration was far from over. My generation ushered in the dawn of technical exploration of unseen places. We would turn inward to map the human genome, find the Higgs boson elementary particle, clone Dolly the Sheep, manufacture stem cells, and travel into the planet to uncover the mysteries of inner space. We could learn about significant events on the internet and watch real-time satellite images of a harbinger of global climate change. We were dreamers that could go to the ends of the world to explore previously inaccessible places above, below, and inside the ice.

THE NIGHT BEFORE the pitch meeting with National Geographic, we had a destination and a historical mentor, but no solid story beyond retracing Shackleton's journey. But as we struggled to find a narrative arc, it was delivered to us on the internet. We had been watching satellite photos that showed cracks developing in the ice sheet near where Shackleton had once camped. These cracks were growing bigger and connecting with other rifts in the ice. Then it happened. The biggest iceberg in recorded history broke away from the Ross Ice Shelf. It was now the largest moving object on the planet, and we decided we were going to intercept it and be the first people to dive inside caves inside an iceberg. Our science team would help us describe the biological wake of the giant berg while we figured out how to bring back images from inside.

But even after National Geographic signed on, there was still plenty to do. We had no scientific permits, no plan, and no idea whether we would even find iceberg caves. It was merely a hypothesis based on our understanding of how caves were formed in limestone. We guessed that if cracks and fissures were eroded by water to form caves in rock, then surely the same process would happen in ice, even faster. Were we certain? No. Would it be easy? No. Would it be safe? No. But that's why it's called exploration. If it were a sure thing, there would be no point in going.

In 1958, representatives from twelve countries gathered in Washington, D.C., to begin talks that would result in the Antarctic Treaty, an agreement that would usher in an era of protection and scientific collaboration at the bottom of the world. All parties, including the United States and New Zealand, agreed that scientific activities would be conducted under permits and that observations and results would be shared with member nations. Antarctica would be sliced like a pie, allowing countries to control activities within their wedge and ensure that scientific projects could happen without disruption from other nations. It was later agreed that Antarctica would not be used for resource extraction and that wildlife and environments would be protected from harvesting. The American McMurdo Station offered the best resources on the continent. Each year, a small town full of scientists and support personnel wintered over in relative comfort to do their research. They had aircraft and ship support during the summer months and all the comforts of life, including a rugby league, churches, and a coffee house.

Now that we had National Geographic's support, our first task was to convince the U.S. National Science Foundation, or NSF, that we had a worthy project so that we could secure a permit. We assumed it would simply be a matter of paperwork to endorse our grand vision.

Dr. Greg Stone of the New England Aquarium joined the project as chief scientific officer. A specialist in whales, he had conducted research in Antarctica before and knew his way around the permit ropes. As the author of our project's *National Geographic* article and principal contributor to scientific papers emerging from our mission, he would take the lead in getting us a permit.

My job as the diving safety officer was to provide the operational plan that detailed the diving technology and procedures. Because I was the only team member with a solid background in cold-water diving, I would be responsible for equipment choices, team training, and supervision of the plan. With plenty of cold-water ice dives under my belt from Canada, I knew I would not feel safe conducting penetration dives into icebergs on traditional open-circuit scuba gear, and so the first decision I made was to insist on the use of closed-circuit mixed-gas rebreathers— the very ones we had used at Wakulla. As they had before, they would increase our range with fewer tanks. More importantly, the scrubber's chemical reaction would offer some warmth, and they were also far less likely than standard scuba to result in a catastrophic gas loss caused by near-freezing water. In cold water, the valves inside scuba regulators can easily freeze open, dumping your entire cylinder of gas. This won't happen in a rebreather. In my mind, rebreathers were the only option. It was a matter of life or death.

Unfortunately, rebreathers were still relatively new to the scientific diving community, and my friends and colleagues were dying at an alarming rate while using them. The NSF responded to our permit request by saying they would have no part in a project using rebreathers. This was a real setback: months of effort hadn't paid off, and we wondered whether there was any way forward. Greg and I discussed the options and I stood firm on using rebreathers. Then the United States Coast Guard sent a communiqué. As a result of the NSF rejection, they would not offer rescue support if we called on them to help. That was it: access denied. No permit or support would be available from the U.S. government, period.

After more endless negotiations and paperwork, Greg, a dual U.S. and New Zealand citizen, landed our permit a month later, under the Kiwi flag. They did not have the assets to support or rescue us, but they were happy to offer us the endorsement that we needed to head toward the pole.

Going to Antarctica is like going to another planet, one that has to be protected. The Madrid Protocol of 1991 established permit and exploration rules and was the first step toward permanent protection of the Antarctic environment. The major conservation initiative of this agreement was to ban mineral exploitation for at least fifty years and to set down measures regarding waste management, marine pollution, and preservation of flora and fauna. It created a system of protected areas and required operations to examine and declare their environmental impact. Because we could unintentionally transport viruses and bacteria to the pristine continent, we would have to decontaminate our boots every time we got on and off the ship. We would need to carefully collect every food scrap from plates and the sink trap and stash the organics in an empty fuel drum until we could carry them home. Some foods were strictly prohibited by the Antarctic Treaty, such as chicken because avian diseases carried in the meat could decimate entire populations of endemic birds. Our permit was strict: it specified the islands we were allowed to visit, whether we could launch a Zodiac boat there, helicopter flight times and zones, and even whether and how we might set foot on land or ice. All these rules were aimed at leaving the continent unsullied, the marine life untouched, and research efforts undisturbed. It was a formidable responsibility to head into such a pristine environment and do no harm.

WE HAD TO carefully consider the risk of decompression illness. With the nearest available recompression chamber located in New Zealand, nearly 2500 miles or two weeks' boat travel away, there would be no hope for proper treatment if one of us got bent in Antarctica. And our boat didn't have the space or infrastructure to house a hyperbaric chamber on board. Dives would have to be planned conservatively—if you could call anything in Antarctica conservative.

But diving was only one of the perils we had to assess as we planned the project. There were hazards that were simply impossible to anticipate.

We prepared and packed the unpretentious vessel *Braveheart* in thirty-five-mile-an-hour gusts that were battering the port of Wellington, New Zealand. Our red two-seater Bell JetRanger helicopter was disassembled and tied to mounting hooks on the upper deck. The rotor blades were carefully packed into shipping crates to make the trip south. Three enormous horizontal tanks for diesel fuel were craned aboard and affixed to the ship's rails with massive ratchet straps. Our plan was to use the fuel in these deck tanks, dropping any emptied containers on a subantarctic island and retrieving them on the way home. We would keep one empty tank to use as a trash and sewage receptacle, since we had to take everything out with us. Remaining deck space was shrinking fast, and everything had to be secure. Captain Nigel Jolly described *Braveheart* to us as a "half-boat/half-submarine" type of vessel. He said that the rear deck would have up to four feet of water on it from periodic waves crashing over the ship. But the water would run along the deck and drain through the scuppers. I imagined beach boy Wes surfing from the fantail to the bow while howling with excitement.

A truck pulled up with our food order, and eighteen of us made a human chain to load big white pumpkins, racks of lamb, coffee beans, canned food, and cases of beer into the forward storage hold. The ship's cook, Carol, got busy chopping vegetables into large chunks and bagging them for the chest freezer that would only need to be powered for a few days. By the time we got below the Antarctic Circle, it would be cold enough to use the whole deck as a freezer. But even the six thousand calories per person per day she budgeted would leave us withering and hungry by the voyage's end.

Captain Nigel carefully measured every drop of the fuel delivery that settled the boat deeper into the wharf. His grown son and second mate, Matt, gleefully loaded his precious snowboard into the wheelhouse. Wes tinkered with his brand-new HD camera. It was only the sixth camera of this model off the Sony assembly line, and we were going to be shooting in one of the most difficult landscapes imaginable—white on white. Paul and I stowed the rebreathers carefully in a tiny, secure room on deck. Everything was tied down and tucked away for the big seas ahead. The air was thick with excitement and trepidation. We were casting ourselves into the unknown with a fragile connection to the world through a satellite phone. A Russian icebreaker three times our size was already steaming its way south with a passenger list of wealthy tourists. We would soon follow them toward the pole.

We had barely pushed off the pier in Windy Wellington when a punishing gust of wind slammed us forward into the dock and the *Braveheart* clipped a loading crane, leaving a deep dent in the bow. The crew scrambled to regain control of the vessel while I tried to stay out of the way. It was a scary start for our journey and a taste of the challenging weather ahead.

As we struck the dock with that deafening thud, Paul let out his first volley of puke. Prone to seasickness, he crawled into his bunk in the companionway of the small vessel and hunkered down under the covers. "This is going to kill me," he groaned. I gave him a Dramamine pill and acu-pressure wristbands, then cranked up his favorite tunes on his Walkman, hoping the music would help him rest and perhaps take his mind off the brisk wind and growing seas. I felt terrible for him, but there was little more I could do to make him comfortable. It was only a short day trip to reach Lyttelton on New Zealand's South Island, where we would complete our final provisioning before hitting the open ocean. Surely he could manage that.

Savvy travel operators were discovering that there was a demand for tourism in Antarctica. The first luxury cruise liner visited the continent in 1966, carrying fewer than a hundred people. By 1990, nearly five thousand people were visiting each year. Cruise companies introduced even larger vessels, bringing twelve thousand visitors in the year 2000. Yet most tour companies made a relatively short twenty-four-hour crossing to the Antarctic by leaving from South America. The seas of the Drake Passage, between South America and the Antarctic Peninsula, can be formidable, with wave heights reaching a monstrous hundred feet, but it is generally only an overnight crossing. And most tourists stay at the warmest part of the peninsula, never going as far as the actual Antarctic Circle at 66° south. Even fewer travel to the Ross Sea from New Zealand during the extremely short and unpredictable summer season. So when people inform me that they have been to Antarctica, I chuckle a bit under my breath. There is simply no comparison between a carefully managed tourist experience and the real threats and discomfort we endured on our crossing.

To reach Antarctica from New Zealand, you must cross the Southern Ocean and traverse the latitudes of intense westerly winds. The ocean area from latitude 40° south to the Antarctic Circle and beyond has the strongest average winds on the planet. Massive cyclonic depression systems spiral toward the continent and establish a permanent ring of low pressure known as the circumpolar trough. At any time, there may be a half-dozen or more fronts closing in on the continent, causing some of the roughest seas imaginable. These latitudes are known to mariners as the Roaring Forties, the Furious Fifties, and the Screaming Sixties. They have propelled brave yachtsmen on round-the-world voyages on storms that ring the planet without interruption. And it is understood that whatever you experience in the Roaring Forties will only get worse as you head farther south. So our intrepid crew of eighteen was braced for a beating.

Within minutes of our evening departure, I knew our first night would be a sea trial for the *Braveheart*. The waves were so violent that I could barely walk around the 118-foot-long boat. A trip to the ship's head required opening heavy steel latches on the waterproof door, then briefly slipping along the hardwood deck to reach a second thick bulkhead. Once I was inside the head, the unwieldy door had to be clamped shut and I had to balance and shift my weight to keep from being thrown off my perch on the rim of the cold stainless steel toilet bowl. Just using the bathroom was going to be dangerous. I crawled into my top bunk and tried to sleep to the vibration of the engine and Paul's seasick moans below me.

Mere hours into the journey, a thirty-foot wall of water hit the beam of the boat, sending everything tumbling. Even boxes placed flat on the floor spilled their contents and crashed up against the wall. I came down from my bunk and tried to clean up some of Paul's barf bucket that had just tipped in the aisle. I climbed up the companionway steps, but the ship kept dropping out from below me. My legs felt like useless rubber bands and my stomach was rising up in my throat. The galley portholes offered a terrifying submariner's view of the undersea world—it was like looking in the window of a front-loading washing machine. Water was injected through the scuppers like a fire hydrant, and the decks were awash with foamy seas. I wanted to reach the wheelhouse to see if everything was okay, but decided it was safer to retreat back to my bunk. Paul was lying in the fetal position, his back to the wall and arms braced against the wooden ridge on the edge of the bunk. I grasped his hand for a moment in an attempt to offer strength, then climbed into the upper bunk.

Sleeping was going to be challenging, especially when I was nearly falling out of my berth on the heel of every wave. I got some rope from my bag and tied a lattice of macramé between the bunk and some ceiling hooks, leaving only a small hole for me to crawl through. My web would contain me in a coffin-sized space for a good portion of the twelve-day transit to the Ross Sea. With my shoulders pinned to the flimsy mattress, I was able to press the palms of my hands on the ceiling to brace myself in the growing torment of the ocean. We were only hours into the expedition, and I was already cut and bruised from simply trying to stay in my bed.

While the boat tossed and turned, I had lots of time to think. I wondered about the emotional gulf that separated me from my husband in the bunk below. I had once felt so strongly about him that I had stood at an altar and pledged the rest of my life to him. Yet now, I wasn't sure whether I knew him or even knew myself. In my family's view, women should be married, start families, and live happily ever after, and men should have respectable jobs that they stay in for life. I didn't do any of that successfully. I might have rushed to the altar partly because of societal pressures, but I felt that if I divorced, it would be the greatest failure of my life. But, I was no longer willing to acquiesce and try to be somebody that just made Paul happy; I had to make Jill happy, too. That meant being honest about what I wanted. I was realizing that I had been enduring a marriage that wasn't fulfilling for either of us—but it's terrifying to start over, and sickening to think that I might hurt Paul. I wanted to put myself first, but I kept hearing the word "selfish" in my head. If I divorced Paul, would that be selfish? I still had some serious decisions to make.

By dawn, the seas were following us with an average ten-foot height, and many of us were able to get up and get a little work done. Paul was still unwell, and not interested in much more than a few sips of water. As the day drew on, the waves and wind increased to a howling sixty-five miles per hour. The boat tilted slightly, forcing us to walk with one hand braced against a wall. The occasional rogue wave would send everything pitching, and one by one, we'd all returned to our bunks. I tried to type journal posts on my laptop, but it just made me dizzy and nauseated.

Finally we turned the corner and entered the haven of Lyttelton Harbour. And just like that, as if someone flipped a switch, the waves disappeared. Rolling green hills plunged steeply into the ocean, and the scent of wildflowers filled the air. This historically significant port had served as the jumping-off point for most of the notable Antarctic expeditions: Sir Robert Falcon Scott, Sir Ernest Shackleton, Roald Amundsen, and others all began their expeditions from this protected anchorage. In all their accounts, Lyttelton was remembered as the last refuge with the comforts of home and a welcoming oasis after a grueling voyage.

We topped up the ship's fuel tanks to their maximum capacity, and the crew bid farewell to their families with last-minute phone calls that left everyone feeling a little melancholy. Before leaving for New Zealand, I had written letters to my parents, brother, and sister, making sure that, in the event that we didn't return from our expedition, nothing was left unsaid. A phone call from Lyttelton would have been too difficult, especially because I was feeling so anxious about the journey. But for once it wasn't the dives that worried me but, rather, everything else: I was about to cast off for a twelve-day crossing over the roughest seas on earth. My husband was barely functioning after only twenty-four hours, and I could see that he was not as eager as I was to get to the southern continent. And my marriage felt on the brink of ending.

A skilled local harbor pilot eased our vessel between Godley and Adderly Heads, over a volcanic sea-filled crater that had not erupted in eleven million years. The pilot successfully drove out of the harbor, then wished us a safe passage, jumped aboard his small tug, and went on his way. There was no turning back now. I popped a seasickness pill. It felt like we were about to drop off the edge of the earth.

For three days, the seas battered us and rendered the expedition team completely useless. We were confined to our bunks to avoid injury, and any ideas about working were washed away. I continued to bring Paul water, and each time I'd take away smaller and smaller buckets of vomit. He was becoming severely dehydrated, and no amount of drugs, wristbands, headphones, and sleeping pills alleviated the agony he was experiencing.

One particularly rough night, one of our fuel drums broke loose on the aft deck. The webbed ratchet straps, now soaked with seawater, had stretched, and the heavy drum broke free and started sliding across the deck. The shifting weight was troublesome enough, but then the drum began spilling valuable fuel on the deck. Anyone who could sprang to action to get it tied down again. I managed to crawl down from my bunk and reach the doorway, but was told to stand by as Wes and the first mate took charge—they had the gear and solid stomachs to run out onto the slippery deck and try to regain control of the drum. The captain spun the boat around and idled in the opposite direction so that the crashing seas would be more manageable—it was better to move with the waves than to have them come smashing down on the bow. Unfortunately, it also meant that we were now sailing in the wrong direction.

For a full two hours, Wes and the first mate scrambled to get the drum under control. I felt useless watching on, but knew that I had an important role in keeping track of every person on the deck and watching for injuries—the giant drum was like a pinball on the deck and could easily crush someone against the rail or knock them overboard. Finally, chilled to the bone and exhausted, they finished securing the tank and collapsed inside the galley. Crisis averted for now, but it would not be the last emergency that required mustering all hands.

It was four days into our journey when Paul reached his limit, or rather, "retched" his limit. The winds had been growing stronger each day until, finally, the seas piled to lofty sixty-foot peaks. Captain Nigel Jolly and Master Iain Kerr rotated shifts at the helm. Nigel's biceps tensed under his pilled argyle sweater as he wrestled the ship's wheel into submission. His sharp features and wild silvery blond mane conjured visions of Viking explorers, confident and in command of the wayward ship. We climbed the steep face of each wave and hesitated slightly as he forced the boat in line before it plummeted down the far side into the trough below. At times, waves sledgehammered us completely underwater, and the ship rolled and pitched then shuddered to pop to the surface again. Dishes crashed from their secure cupboards, equipment fell from our bunks, and worktables and everything that wasn't tied down bounced around the boat. This went on and on, relentlessly, through the darkest night I can recall.

I was wedged in my bunk between sleeping bags and down outerwear when Paul called out for me. Over the last day, he'd been reduced to gagging and groaning, unable to keep down even the smallest cup of water. He was getting weaker and was thinning by the minute. I toppled out of my bunk and took the feeble hand he extended to me.

"How can I help?" I asked with worry.

"You lied to me!" he announced, clutching my hand with remarkable force.

"What do you mean?" I timidly asked.

"You said this was a twelve-day crossing!"

I stuttered, "It is. It *is* a twelve-day crossing, sweetie."

"A twelve-day crossing is six days down and six days back!" he said firmly. He had thought he was almost over his agony when some well-meaning crewman told him that more than a week of this lay ahead.

"No, darling," I said gently. "A twelve-day crossing is twelve days down and twelve days back."

"That's it. I'm going to die here," he gasped. "It's all your fault. I will never make it home."

I felt terrible. I never imagined that this part of the journey would be so hard; I had been optimistic, thinking the weather would calm or Paul would adjust. But the physical and emotional stress were taking me to my breaking point. I could no longer fight the conditions either, and began to retch into Paul's bucket. I still wanted to get to Antarctica, but I knew that Paul would have jumped at any chance to go home. But the reality was that we were going to have to survive this ordeal for the next two months.

When you are tired and uncomfortable, closely held feelings rise to the surface, and it's impossible to ignore them. With more than a week of big seas ahead and barely able to move from my tiny bunk, I had nothing but time to think. I felt guilty about asking Paul to come to Antarctica when I no longer felt the pull of attraction that had first brought us together. But how could I have gone away for two months without him? No matter how hard I tried to imagine us as a couple, I just couldn't see a happy future. And when I looked ahead and saw myself alone, I felt a sense of relief. There were many uncertainties—where would I live, or how might my career play out?—but those felt less and less important than releasing us both from what felt like a performance rather than passion.

THE BOAT, THE team, and the crew were completely battered—and it showed in our sallow faces and weak bodies. Finally, the captain decided to take a detour, to give us all a break. We sought the last land refuge available, Campbell Island, at 52° south. That night, I felt the turn of the vessel and then quiet calmness as we motored into a harbor protected from the wind and waves. Even though it was past midnight, I was just so thrilled to be able to rise safely from my bunk that I went up on deck with the crew. Although we could only see a small patch of flourishing green illuminated through the foggy spotlight, it was a beautiful sight: a single Sitka spruce, the last one on our way south. Even Paul managed to get out on deck for the celebration that ensued. Chef Carol started cooking and the crew passed around some Tui beers. None of us had eaten much in the previous days, so we gorged on lamb shanks, pumpkin, and mint sauce.

The next morning the weather was moody, a moment of sun being snatched away by ominous black clouds that brought squalls with gusts up to 100 miles an hour. Our plan for the day was to stretch our legs and take stock of any damage to the vessel and our gear. I joined our science team in the Zodiac inflatable for a wild-life survey, and they showed me rare yellow-eyed penguins, flocks of sooty albatross, royal albatross, sparrows, gulls, skuas, cormorants, and nesting Antarctic terns. I wondered how these beautiful birds survived the harsh remoteness of this deserted island group, because with the howling wind, tall tussock grass, bristly bushes, and blasting sand, I couldn't imagine a tougher place to live. We walked along a beach populated by deafening elephant seals and Hooker's sea lions that were larger than our skiff. The island felt ancient to me, and for the first time in my life I did not feel like the dominant species. A large elephant seal bull lunged at me, baring his teeth, when I got too close with my camera. It was a terrifying reminder to me that this wasn't just our planet.

While we were exploring the island, back on the ship the crew uncovered a serious problem with our helicopter: a battery that was left connected to a relay had been exposed to the pounding seawater. When a crewman touched the wiring on the belly of the helicopter, the wires crumbled in his hands. The chopper was critical to our mission ahead, needed to scout leads in the ice pack or search for us if we became separated from the boat while diving. Laurie Prouting, our pilot, was brilliant, but he had no experience in repairs, and so we had to carefully preserve and document what was left of the dissolving electrical system. It became a game of tracing wires and soldering replacements in an attempt to copy and rethread what was visibly damaged. The gravity of our situation was weighing on us: our satellite phone had already failed, and we had no contact with the outside world. No phone. No internet. No calling for help. And no helicopter. We were alone against the elements.

As we departed Campbell Island, we managed to establish a nightly radio check with a woman named Meri Leask, of Bluff Fisherman's Radio in New Zealand. The amateur voice of consolation would try to reach us each night at 9:00 p.m. It was comforting to know that someone was looking out for us. Meri maintained contact with other ships in the Ross Sea and would share weather observations that she got from other captains. She could also serve as an emergency relay if we needed help. The second point of contact we made was a woman named Belinda who was aboard a Russian icebreaker turned cruise ship that was also heading south. The *Kapitan Khlebnikov* was currently a day behind us but would soon overtake us. (We were traveling at a slower speed to conserve fuel.) Although I was disappointed that a tourist cruise ship sailed these waters, it was a godsend to have someone just ahead of us warning of obstacles like floating pack ice or heavy seas. The *Khlebnikov* had an ice-strengthened hull, but our delicate boat was as fragile as the *Titanic*. I worried constantly that we'd strike unseen obstacles of ice jutting out from bergs.

Some of our crossing brought relatively benign wave heights in the range of fifteen to twenty feet, but after leaving Campbell Island, we were racing between two storm fronts that were closing in on us and expected to merge. One night, we were jolted awake by the largest wave we'd experienced so far. The towering forty-footer slammed our starboard side with a sound like we had hit a rock or, worse, another ship. Everything crashed to the floor, and things on the floor went airborne, and all we could do was endure the battering. Petrified, and my muscles and mind exhausted, I braced myself between the wall and piles of clothing on my bunk. It turned out the ship's inclinometer recorded an estimated fifty-degree sideways roll. John, the ship's second engineer, said this was a new record for him. The needle on the inclinometer was now stuck in place, pegged beyond the end of the instrument markings for the rest of the trip.

"Why do the markings stop at forty-five degrees?" I naively asked John.

"Because you don't usually recover from anything worse," he replied.

I felt a chill run the length of my body down to my toes. We'd been lucky.

The worse the weather got, the more informal meals became. When she was able to cook, Chef Carol would prepare something and leave it hog-tied to the stove with a spill-proof lid lashed to steel rails around the burners. If you managed to get to the galley and spoon out a bowl of thick stew, you had to eat it fast or you'd be wearing it. I discovered that if I stood in a lunging yoga pose and swayed with the seas, I could get some food in my mouth without too much mess. Even then, though, it was clear very early on that keeping clothes clean would be impossible. Every meal rewarded me with new souvenirs down the fronts of my small ration of shirts.

Rime ice from sea spray would coat the upper reaches of the vessel, which was always dangerous—weight on the boat's structure above the waterline risked overturning the ship—so we would be gathered on deck, given baseball bats and hammers, and instructed to remove it. Sometimes it felt like we were riding a raging bull in the middle of an ice storm, the ship moving so violently we'd have to secure ourselves to a rail just to smash the stubborn ice. I could feel the adrenaline coursing through me. Below deck I was hostage to the elements. On deck with an aluminum baseball bat, I could at least fight back.

A week into the trip and close to midnight, the captain announced that we had reached the Antarctic Convergence zone, and he made a slight course change to handle the waves a little better. In this twenty-five-mile-wide belt of mixing water, between 47° and 63° south, cold north-flowing currents meet the warmer circulating waters, and the water's surface temperature and salinity change dramatically. We watched as flocks of birds dived into roiling schools of fish at this biological demarcation zone. At this intersection of the subantarctic and Antarctic regions, different species of birds and fish live on either side of the convergence. Our ornithologist, Porter Turnbull, pointed out wandering albatross and sea petrels as he excitedly made notes in his logbook. Their giant wingspans cast a shadow as they soared overhead, offering me a touch of optimism. If these birds could live here, surely I could manage a couple of months.

It was the perfect opportunity to do some filming, so even though Wes, Paul, and I still felt weak, we did our best to summon our energy. We filmed our science team—Porter, chief scientist and whale specialist Greg Stone, and marine mammal specialist Carlos Olavarría—out on deck. They were giddy from all the bird sightings and the whales swimming near our boat.

We attempted to get some preliminary interview segments with each scientist, but it became almost comical trying to keep them in the camera's frame because the ship was still lurching like a wild horse. Greg Stone sat in front of a small wooden rack of books in the forecastle. Old maps were strewn across the desk, and brass instruments decorated the walls. Chef Carol was put to work as a human sandbag. She sat cross-legged beneath the camera tripod and used her weight to try to secure it firmly on the floor. I asked Greg to describe the journey. He laughed with caustic nervousness and described the terrors of the previous week. I asked him to talk about his work as a specialist in whales, but just as he began to answer, a massive wave hit the ship, and we all slammed against the hull. I threw my body between the wall and the camera to cushion it from damage, and eruptions of laughter came from the pile of bodies and gear. It was utterly ridiculous for us to try to work, but it felt good to do something productive. Wasting away in the bunk is tough on the body and mind.

OUR JOURNEY TICKED onward, passing latitude landmarks that eased us closer to the unknown adventures ahead, including the imaginary line that marked the start of Antarctic Treaty waters. This was where our hard-won permit kicked in. After nine days at sea and nearly 1900 miles traveled, we crossed the next indiscernible border, the Antarctic Circle. British Navy captain James Cook was the first person in history to circumnavigate the Southern Ocean and cross the Antarctic Circle, between 1772 and 1775. While this feat sparked a rush of enterprising whalers to the area, Cook maintained a less optimistic view of the merit of his travels. He could not imagine that the promise of good hunting was worth the risk. In his journals he wrote, "I can be bold enough to say that no man will ever venture farther than I have done; and that the lands which may lie to the south will never be explored." He went on to say, "If any one should have resolution and perseverance to clear up this point by proceeding farther than I have done, I shall not envy him the honour of the discovery; but I will be bold to say, that the world will not be benefited by it."

For us, passing the Antarctic Circle was something to celebrate. We were getting closer to our ultimate destination and the seas were becoming tamer—or, rather, I was at least getting accustomed to them. Nigel led the group in a bit of a party that involved a ritual of drinking rum and spouting silly recitations. After this, we were christened Antarctic explorers, and Master Iain presented us with ties and woolen scarves with the rare Antarctic tartan. We got tipsy on the rum and then spotted our first iceberg just beyond the Balleny Islands. It was the first tangible evidence that we were finally getting close. At first it was just a fleck on the radar, then a tiny bump on the horizon. When we finally drew close enough to see it, I was nearly moved to tears. The glittery surface of the towering mountain of ice looked so magnificent that I felt like the goal of our voyage was finally within reach. At last, there was an end to the stresses of travel—and that meant we could begin our expedition.

Later that evening we radioed Belinda on the *Kapitan Khlebnikov* to find out what lay ahead. She described a grim outlook: that year's pack ice was the very worst that anyone could recall. Her vessel had been unable to reach land at Cape Adare or Cape Hallett on the Ross Sea. A single icebreaker had managed to make it safely to McMurdo Station to resupply the American base, but the sea soon iced over and trapped the boat, and it seemed unlikely that any other supply ships would reach the bases this season. Despite an invitation, we didn't have enough fuel or open water to reach New Zealand's Scott Base and connect with the scientists there. Too much ice separated us from its location to the south.

The new challenge we had to face was that we were running through our fuel supplies. The tough seas and high winds had required us to burn more diesel than we could have predicted. So, with no options for refueling (we couldn't just power down the ship and expect to stand in place—particularly when we were fighting currents and winds), and with the clock ticking on finding our iceberg caves, we had to make some changes to our route. Wes, Greg, Iain, Nigel, and I met in the wheelhouse to discuss the options. On the nautical chart, Iain pointed to our current location. Based on his estimations as to where we would find the B-15 iceberg, we didn't have any time to waste. Instead of taking a meandering course via Cape Adare, where we could have made landfall, reassembled the helicopter, and slowly eased into diving, we'd have to head straight toward B-15, assembling the chopper and preparing to dive in transit. If we did not hurry, we'd risk getting trapped inside the ice as the Antarctic winter set upon us. Fuel shortage was not an issue I had anticipated, and so it made me incredibly nervous to hear the concern in the captain's voice. Every time I saw him check the fuel supplies and tally the totals, it made me even tenser about getting home safely. We turned down the thermostat to the temperature of a meat locker and slowed our speed to conserve every drop of fuel possible. I tried not to worry, and instead, from that day forward, never removed my hat and gloves and wore my Canada Goose down parka inside the ship day and night.

In 1915, Sir Ernest Shackleton and his crew aboard *Endurance* were trapped in the pack ice for 326 days. As the Antarctic spring began to shift the ice around them, they looked forward to finally being released from their frozen prison. But just when things seemed to be improving, the unpredictable ice savagely crushed the vessel instead of freeing it. As the pack ice fell apart and their boat began sinking, the formerly hopeful men had no choice but to salvage as many survival supplies as possible and abandon the ship. They got onto a shrinking floe and established Patience Camp. As they drifted north at the whims of the ocean, days turned to months. Their depressing uncertainty and heroic struggle were only just beginning.

In our new reality of dwindling fuel supplies, we knew we needed to find our own Patience Camp—a place where we could work out diving and filming protocols and get the chopper safely in the air. We needed a place where we could turn off the ship's engines and save fuel—and we needed to do it quickly if we wanted to avoid having to turn back for New Zealand without a single

frame of our film shot. We also had a full roster of scientific observations yet to make. We were not ready to accept defeat.

Now that we were well beyond the Antarctic Circle, the pitching and rolling began to subside, and we inched forward into the
endless white pack ice. The sun was no longer setting, and I wasn't
the only one who had a difficult time sleeping with the constant
daylight. And the stress of unmet objectives was weighing heavily
on me. We had bet our reputations on delivering footage from
underwater caves and we were barely entering the realm of the
pack ice—what kind of expedition would we have? The thick ice
made it challenging to move toward B-15, and every time we hit a
piece of ice, it loudly ground and thumped against the hull as
though it was tearing us apart. I hadn't imagined that we would
be in such close contact with the ice, and I was desperate to get
into the water, to see the damage to the boat. I couldn't get out of
my mind those images of the *Titanic*, with a gaping hole in its side,
sinking into the icy sea.

Each day the ice fields expanded until the only way we could
move forward was by following labyrinthine leads of open
water in the matrix of ice. Laurie, our chopper pilot, decided it
was time to reassemble the helicopter so we could see if there
was any open water ahead. He courageously insisted on flying
alone in case our repair work at Campbell Island failed, and I
felt a palpable relief when he was able to alight and get a look
at our situation from above.

I eagerly jumped in the chopper for the second flight and was
shocked when I looked down at the *Braveheart*. From eight hundred feet above the deck, the boat was invisible unless you looked
closely, and everywhere around us was a limitless white landscape
that looked like somebody had smashed white glass with a
hammer. The cracks and gaps extended beyond the horizon of
almost solid ice, with a few narrow rivers of dark open water. The
mountainous bergs that punctuated the ice were only visible

because the low sun lit their edges in a blaze of orange. For the first time in my life, I felt the immensity of complete isolation. Underwater, I was in my element, but flying over this vast landscape, we were barely a flea on the back of this giant planet, with nobody to call for help.

Laurie and I spotted a patch of open water a few miles beyond the boat. If we could escape this thick band of ice, we might have a chance of getting closer to B-15.

The next morning, nearly two thousand miles from the coast of New Zealand, we had to assemble the crew on deck to pry some large floes out of the way. We dropped the Zodiacs in the water to attempt to move the massive chunks of ice, but the ice was too heavy. Tensions were rising when Laurie launched the chopper to guide the vessel forward, picking the best path away from the thickest pack. We wanted to move deeper into the Ross Sea, but we couldn't risk being trapped in the ice.

But getting trapped in the ice is precisely what happened next. The end of the Antarctic summer was fast approaching, and the sea to our south was freezing solid toward us at a rate of one mile each day. I was beginning to think we were going to have to retreat without our story, and then a great blast of wind rushed the ice to fill the open space all around the *Braveheart*. We pushed the engine as hard as we could, trying to get the boat through the pack. Nigel and Iain climbed up the mast to look for a way out of the ice. The slush and lumps rolled and bounced along the hull. Sometimes a large floe was dragged alongside until we slowed the engine to release the floating block. I read the faces of the experienced crew. They were just as terrified as I was.

Once again Laurie put the helicopter in the air and tried to guide us through the ice that sucked in around the vessel, and for nearly two hours Nigel tried alternately backing up and moving the ship forward while attempting to spin on a dime. But the ice was acting like glue on the structure of the ship and it was not going to let us go.

At 2:30 a.m., Nigel hailed all hands to muster on deck, and we all lined up for instructions. I was dressed in my heaviest overalls and parka. Ski goggles, a neck warmer, and a thick wooly hat were snugged tight in an attempt to cover every bit of my exposed flesh. Wes and I were rolling the cameras, but we were ready to set them down and help with what was quickly turning into a real emergency. None of us had rested for a day and a half, and extra hands and moral support were needed. We dropped the Zodiacs in the water—they'd be our bow and stern thrusters—and the helicopter was sent back aloft to search for leads. We took up positions around the rail to relay instructions to the two captains of the inflatables.

The news from the helicopter was grim: we were completely encircled by ice. We could conserve fuel by turning off the engine, but even if we did get released, we were losing ground. The wind was blowing us north with the ice, away from B-15.

Now that we were stuck, the engine went silent for the first time in almost a month, and the absence of sound was unsettling. I had grown accustomed to its white noise. It was the heartbeat of the hull, and it kept me moving through the days. With no darkness to steer my circadian rhythm, I was thrust into a weird sleepless limbo, and I was losing track of hours and days. With no regard for the actual time, I worked until I was exhausted. Although I wanted to eat when I was hungry, there seemed to be nothing that would fill me up. Then Nigel decided to limit mealtimes, fearing that we would run out of food.

The ship and the ice floe drifted back and forth across the International Date Line, so we were losing and gaining days on the calendar until nobody was clear about the actual time or date. But on this day, trapped in the ice, it was gloriously warm and sunny, perhaps the nicest weather we had seen on this trip. We asked ourselves, "What would Shackleton do?" We knew the answer: he would keep morale up—he would emphasize teamwork but also laughter and play. Photographer Frank Hurley's photos from 1915 show Shackleton's men playing soccer on the ice floe, their sinking vessel a backdrop. So, in their honor, we decided to play a game of touch football on the ice, at the same time the Super Bowl game would be happening at home. We placed bets on the Baltimore Ravens or New York Giants, then stripped down to T-shirts and left the boat in the hands of Master Iain. Sweating and tumbling on the ice, we worked out our frustrations and stress, enjoying the moment and abandoning the dark thoughts of what might come next.

I grabbed a camera and asked Wes to stand on the flat ice with his back to the boat. From where I stood, I could clearly see the ice heaving up against the hull. The *Braveheart*'s bow was proudly sticking up into the blue sky, but it looked off balance from the stress of being trapped. It was a great image for the film, but an ominous sign of our dire situation.

Wes was a brilliant storyteller, gifted at encapsulating the moment in a way that would clutch the hearts and minds of our audience—and he could do it on the fly, with minimal prep. As I watched Wes on camera, I was instantly connected to the unfolding emotional drama. "What is happening to the boat behind you?" I asked him. "Are we trapped?"

His face turned grave. "We've followed Shackleton's ghost across the Southern Ocean, and now find ourselves exactly in his shoes, trapped in the ice, perhaps awaiting the moment when our vessel is eventually crushed." I zoomed out to reveal the boat behind. It was an excellent performance that would have terrified anyone— except, while I filmed, second mate Matt Jolly was running back and forth between Wes's back and the boat, completely naked and waving his arms in the air like a lunatic unleashed. I've often noticed that during what should be the peak of stress, people dig deep to find a very stoic and dark humor. If our ship were to sink, perhaps we would save this piece of film, just as Frank Hurley did, so the world could see our final days. Except, rather than the dignified photos of

Shackleton in his woolen sweater, we would have the pale-skinned Yeti, Matt Jolly, entertaining the troops in all his Kiwi glory.

Back on the boat after our football game, we discussed rationing the food. People were getting edgy, and they were all dealing with their stress in different ways. Captain Nigel was acting strangely, seizing and locking up our chocolate bars and snack stash, then dispensing treats only to those he liked. Below deck, our remotely operated underwater vehicle pilot was curled up in the fetal position in his bunk, paralyzed in fear. The chopper pilot was relaxed but idly reading *The World's Worst Aviation Disasters*, and our soundman was three-quarters through his bounteous sixty-day private stash of liquor. And we were all concerned that our urine was a deep brown—the sign of a failing water maker or rusty holding tanks. I kept busy by reviewing and logging our footage, and Paul tinkered with the rebreathers below deck.

Suddenly, a loud rumbling broke through the tense silence of the ship. The ice was moving! I quickly climbed topside and saw that we were being freed from the grip of the pack. The engines were immediately powered up, launching with them a hive of activity. I was incredibly relieved. In the face of danger, I try hard to deny the darkest thoughts about terrible outcomes. Worry can be infectious and toxic, so I try to keep a positive outlook for myself and others. It's not always an easy creed to live by, but once safe, I'm hit with a profound wave of reality. Our vessel could have been crushed, and we could have lost our lives. But I knew I didn't have time to think about that now—there were tasks at hand, things I could control. With the end of the southern summer approaching, we needed to start diving right away to capture story elements for the movie. Beneath the waves, even waves as dangerous and cold as these, I would be in my element. I had a sense of control in the water. It was my place in the world.

After moving away from the band of pack ice that had trapped our boat, we found a glorious mountain of ice a few miles away.

Nearly one hundred feet high, the pointy summit glowed in a pink-ish reflection of the afternoon sun. We estimated that we were just twenty miles from B-15, but stopped to conserve fuel and finally dive. What we called Patience Camp II provided a diving platform with a small double-story ice cave as well as a large, flat area from which to shoot. Leopard seals and migratory birds dropped in for face time in front of the cameras. Whales, crabeater seals, and penguins taunted us from a distance. Snow petrels set up camp on a small corner of the berg. And we finally had a chance to inspect the hull of the boat. On the first dive, I was surprised to discover that only the paint job was damaged, which was a huge relief.

BEING BACK IN the water after our long, tense voyage rejuvenated me, but I quickly learned about new diving risks. Unpredictable vertical currents were extremely dangerous. An iceberg bounces up and down in the swell of the ocean at the same time that fresh meltwater pours down its flanks. When the freshwater, swells, and salt water combine forces, the result is a violent mixing of dif-fering water densities that causes rapid downdrafts and brutal upswings. Downward currents will suck you on an out-of-control elevator ride to the depths of the ocean, and when you're finally released, it is even more dangerous to ride the rising water to the surface. Paul and I had to take care, moving slowly while closely watching our dive gauges. At the first sign of a downward current, we would swim perpendicularly to try to escape it. Despite our plan, though, we sometimes found ourselves suddenly sixty or seventy feet deeper than we had intended.

As we ramped up our diving operations, we encountered other perils. On Wes's first dive, he accidentally flooded his new dry suit, but instead of getting back in the Zodiac right away, he decided to grab some footage with the new camera. Paul and I and two other people had a hard time pulling Wes back into the Zodiac. His suit was filled with about sixty pounds of icy water, making him very heavy and hard to grab. It was like wrestling a giant wet noodle onto the boat. It took hours of rewarming to overcome the hypothermia. The experience scared Wes, and he wasn't sure if he was going to do another dive—the rebreather and camera were overwhelming, and he was ready to give up on diving—at least until Paul and I found a cave that was worthy of exploration.

We also learned that ice could move so quickly and strongly that it could crush us between the pack and the boat's hull. Once, I was waiting for my turn to be lifted from the water onto a wooden platform on a boom, and from there onto *Braveheart*. As I waited, the ice moved in and slammed the hull, and the only way I could avoid being crushed was to dive under the ice, then swim under the boat to the other side, and search there for a spot to surface.

The following week everything came together. As we steamed in the direction of B-15, we found larger and larger sea caves and fissures to explore. The science team worked on their observations of jellyfish and whales. The boat tracked a pod of majestic orcas, and we eased our way into their group, moving along at their speed. As their black and white backs teased the surface, loud blasts of spray erupted one by one from their blowholes, backlit in gold by the low sun. I felt a profound sense of awe witnessing the small family moving almost silently along the surface. Carlos readied his biopsy gun, and after several attempts managed to get a skin sample from a large male, the dart sliding across the whale's back just behind his dorsal fin and gouging a long, black skin fragment. (This first DNA sample taken from an orca in the Southern Ocean would enable the determination of a new subspecies of killer whale unique to Antarctic waters.) We also gathered much smaller sea life specimens, populating an onboard aquarium for the study of jellyfish. We filmed the transparent jellies with a black curtain backdrop so we could see their rainbow-colored hair-like cilia that vibrated in a psychedelic light show.

Storm squalls passed through almost daily, and we witnessed the fury of a katabatic storm rushing unimpeded down the sloping terrain from the South Pole. The wind speeds were around sixty miles an hour. While we hid behind tall bergs to stay out of the stormy weather, we struck several growlers—small chunks of ice—hidden beneath the surface. Twice, the large blocks scratched down the side of the boat and lodged in the propeller with a horrific crunch, effectively paralyzing the boat. Fortunately, both times the ice broke away from the prop and we eased forward again.

The next morning we were surrounded by large, flat bergs just offshore of the Possession Islands. We considered diving, but the swirling eddies and rip currents convinced us otherwise. Laurie set off in the chopper to check ice conditions and declared that the way was clear to Cape Hallett and landfall in Antarctica. We charted a course through areas that had three- to six-foot-diameter clumps of ice floating amidst mostly open water. We named these spots "asteroid belts." We arrived on the tip of Victoria Land a day later. This was a historic spot, where the first winter camp was established on the continent. In 1957, the United States and New Zealand worked together to establish Hallett Station. The station was manned year-round until it was heavily damaged by fire in 1964. Cape Hallett was a specially protected area under the Antarctic Treaty System, but we had a permit to make landfall since no other scientific work was scheduled there at the time. We wove the boat through the heavy ice and headed toward the only piece of real rock we had seen in weeks. I was excited to ramp up our cave-diving efforts, and to see my first big congregation of penguins.

The colony of thousands of young Adélie penguins at Cape Hallett could be seen and smelled on the horizon from miles offshore. The odor was akin to seafood scraps left in a trashcan in the sun for several days, multiplied exponentially and added to shit. Most of the hundred thousand or so parent birds had already gone to sea for feeding, leaving behind the youngsters, who were not yet ready to swim great distances. Some adults were acting as caretakers, but as we entered the stinky bay, we watched shadowy predators attacking and snatching the most vulnerable of the colony. With a lofty four-foot wingspan, skua birds generally eat fish and krill, but when mother penguins have deserted the nest, a diet of eggs and lethargic baby penguins is preferred.

We motored into the hamlet of birds, and to our surprise we spotted four figures waving their arms near a humble ribbed-steel Quonset hut on the shore. I picked up my binoculars to get a closer look and saw several small colorful tents within the colony of penguins. There must be research happening here after all. We anchored and tried calling them on our radio. When we couldn't reach them, we decided to take a Zodiac to the group, whose efforts to get our attention were starting to look more frantic. Our male crew jostled for the job of being in the shore party, imagining that maybe at least one of the other group would, by chance, be a woman. The isolation of the sea was getting to the best of them, and by this point in the voyage I was more of a sibling than an interest of anyone's lust.

The group consisted of three Americans—two women and a man—and one Kiwi woman. Their supposedly four-day mission had started more than three weeks earlier, when they'd been flown in and dropped by a helicopter from McMurdo Station to inventory the remaining equipment and infrastructure at Hallett Base and the penguin rookery, and to also remove every trace of habitation: fuel, hazardous materials, heavy equipment, and the Quonset hut, which was full of food rations and books. But thick pack ice had made it impossible for the pickup vessel to reach them, and for the last three weeks they'd had to fend for themselves. Out of fresh provisions, they ate leftover dehydrated and canned food that they found in the abandoned hut. And out of work to do, they had been playing a couple of board games, including Monopoly, for much of the past three weeks. They were relieved to be rescued and excited about a hot shower and a home-cooked meal. We were just as excited that the food rations in the Quonset hut were now fair game for consumption. With food a little lean on the *Braveheart*, we planned on raiding the hut to supplement our dwindling stock.

It felt strange to set foot on land in Antarctica. My legs were wobbly from the weeks on the boat, and I sensed the ghosts of great explorers like Shackleton, Scott, and Amundsen nearby. A profound reverence for the pristine natural environment mixed with gratitude for having a chance to come ashore on the continent even recently marked on the map as Terra Incognita—unexplored land. On the other hand, I just wanted to run like a gleeful child through the penguin colony. As several curious birds waddled in my direction, I found myself giggling at their comical shuffle.

On the peak of nearby Mount Geoffrey Markham, Paul and I found an old glass jar with a rusty cap containing paper logs and a pencil noting every visitor that had made it to the summit. Seeing something manmade felt out of place, but I could imagine the resident scientists climbing the peak on a day off. We visited the station's supply hut, where I helped myself to a few Hershey chocolate bars in vintage wrappers and bags of hard candy made in the 1960s.

With our group now bigger by four, we shifted into a spell of celebrations. We swapped stories, photographed penguins, and, that evening, we passed around drinks and cigars while Wes initiated a jam session with spoons, harmonica, and guitar. Our four stranded guests were thrilled with the change of venue and slept on the galley floor and benches, while the intoxicated male crew tried to entice the women to their more comfortable bunks. Quickly, our visitors' rescue was turning into a bit of a nightmare for them, and in frustration they wedged the galley door shut so they could get a good night's sleep. It made me angry that our female guests had to be wary of anyone in my group, and like them, I decided to stay sober for the entire trip. I had noticed that one of the male crew would relentlessly pursue our female chef every time he got drunk, which was often. I was glad to have a husband on board and be considered "out of bounds."

In the morning, a Coast Guard vessel arrived to pick up our new friends. They had been trying to get through the ice for almost two weeks and were relieved to find the people they had been dispatched to rescue.

Paul, Wes, and I roped up and struck out over the ice shelf fed from the tongues of the Manhaul and Edisto Glaciers. Brown lumps of Weddell seals lolled around on the ice. Startled penguins hurried away, diving into the water, and petrels played in the air currents over our heads. We made our way toward the towering Admiralty Mountains that seemed to loom beyond our reach. In an environment as naked as Antarctica, it can be impossible to envision the real scale of distance and time. Whether I was under-water or hiking on the ice, the panorama seemed so big that I felt insignificant. As long as I walked, I never seemed to get closer to the steep mountains ahead of me.

I was getting accustomed to the constant excitement that made a day feel like a lifetime. Situations that frightened me in the morning might seem mundane and ordinary by day's end. I did things easily that had seemed impossible only hours earlier. Hiking on the ice shelf, roped together with Paul and Wes, and encountering a sleeping seal was not something I thought I would ever experience. At one in the morning, a fiery red sun burned on the horizon beyond the bow of the boat. In the confu-sion that characterizes time in Antarctica, there was a battle between the horizons. It looked like the sun was setting and rising at the same time.

Back on board *Braveheart*, I stood and looked at the water, as flat as glass, lit with soft pinks and pastel blues. The floes were pillowed with the cover of the previous night's gentle snow. Slow swells lifted and released the floating ice as though the ocean were breathing. It was mesmerizing. There was a deeper chill in the air that reminded me that February was upon us and the Antarctic fall was coming. Slushy grease ice was freezing on the surface in the fading late-season light. The solidifying crystal lattice on the surface grew through the twilight until the sun fought it back in the brightness of morning. During the night, Antarctica responded to the still of daytime with what felt like vengeance and reprisal, reminding us that the weather here could be transitory. The winds blew to over fifty miles an hour.

Time felt nebulous and fleeting. We were obliged to recall the most important rule of exploration: we had to be prepared to work to the very edge of what was possible and know when to turn to head home. That time was very nearly upon us.

WE SPENT THE morning in the Zodiac and located an enormous flat topped berg that was likely an orphaned piece of B-15, that was already breaking up into many smaller pieces. We didn't have access to satellite images to confirm our hypothesis, but it was a good guess. And with the majority of B-15—which was estimated to be the size of Jamaica—trapped in the pack to our south, we decided it was time to look inside the bergs on our doorstep. The one we had chosen had an immense crevasse traversing its height, and if we could maneuver ourselves directly into that split in the ice, penetrating the interior of the iceberg as far as possible and descending from that spot, we'd have our best shot at finding more underwater caves.

We were dressed fat in the latest in protective clothing. I bundled up base layers under my dry suit to provide as much warmth as possible, and plugged in a small battery-operated heating pad over my lower back to warm my core. It was an extra twenty pounds but well worth the bulk, because the water was 28°F, just one-tenth of a degree away from the freezing point of salt water. Without that protective insulation, no human being could survive for more than a couple of minutes. Yet, even with our advanced diving gear, the amount of time we could safely spend in the water was extremely limited, and I was nervous. Until that trip, I'd tried to restrict my dives in icy water to around thirty minutes, but now we were planning to be down for an hour, maybe a bit longer for decompression. The risk was clear: any longer than that and we'd be marching toward hypothermia. To make things worse, the temperature of the air outside was significantly lower than that of the water. That meant any time spent floating on the surface was even more dangerous than being submerged. If we separated from the boat, just waiting on the surface could kill us.

Paul and I readied to make the first plunge alone, with Wes skipping the dive. His earlier suit flood had left him troubled, and he wanted to be sure that Paul and I found something significant before he jumped back into diving. We clumsily climbed into the Zodiac, and Wes maneuvered us into position in the middle of a wide fissure in the berg. With a wooden paddle we pushed floating pancakes of ice out of the way so we could safely do a back roll into the water with our cameras.

As soon as I hit the water, intense pain assaulted me. The body-wide ice cream headache made me clamp down my eyelids and gulp. We descended through the slush, and when I opened my eyes, I could see chunks of ice the size of cantaloupes drifting by, but I couldn't seem to focus or orient myself. I flushed out my mask with the frigid water and gasped as I tried to resist the reflex to hyperventilate. The slush, freshwater, and salt water combined to make a slurry through which I couldn't see, and the stinging cold made my vision blurry. As I tried to move the ice out of the way, I couldn't help but wonder what might be below us. A leopard seal, a shark, or any number of predators could be swimming close by.

To reassure myself, I pushed my camera deeper through the slush and looked around. The water was shifting from white to blue, the colors dissolving into shadows at the edges. The shock of the cold was easing slightly, but I could still feel a burning chill right down to my toes. I dropped first through the blurry transition zone to see a beautiful wide crevasse descending out of my sight. The deep fissure seemed to glow with sunlight captured within the undulating surface of ice. Below me was only blackness. The walls of the berg were dimpled, and the eyes of thumb-sized see-through fish reflected my cave light as they flitted around, seemingly surprised by visitors. They darted for their burrows in the ice wall, but their eyes remained trained on us as we passed to go deeper. I began wondering about the cleft we were in. Was it spreading farther apart? Could the changing tides slam it shut? I quickly rid my mind of the negative thoughts and focused on the magnificence of the crystal palace I was swimming through.

As I descended into the dark, I turned on my large canister light, but it did little to illuminate the gigantic space. It was completely silent, save for the occasional click and hiss of my rebreather. Though I was thoroughly enraptured by the majesty of this new environment, I regularly checked my handsets, reassessing the time and depth, decompression time or TTS—time to surface— and the partial pressure or concentration of oxygen that I was breathing. In the corner of my mask, my heads-up display—a row of red, amber, and green lights—summarized my life-support status. The coded flashes indicated that everything in the rebreather was working fine.

With Paul on my left and slightly behind me, I looked to my right to see a large tunnel disappearing farther into the berg and what seemed like a multicolored shag carpet rising to meet us. We had reached the seafloor at 130 feet. I glanced down to my wrist computer and noted that it had taken fifteen minutes to get to this point. Now I could see the iceberg was manacled to the ocean bottom by thick columns, leaving an exposed passage about five feet high.

The underbelly was completely free for us to explore. It was too good to be true. This was exactly what I had hoped to find.

Beneath the turquoise ceiling, the floor of the cave was unlike anything I'd ever seen: a spectacular carpet of densely packed living creatures in vivid warm colors. Vibrant red and orange lumpy sponges, fluttering filter feeders, and other unfamiliar forms of life covered the bottom. Massive cockroach-like isopods swam among the stalks of bottom-dwelling life that looked like wheat fields vibrating in a breeze. There were spiky arrow crabs with black and orange pinstripes, their jointed toothpick legs plucking their way across the ground. It was an ecosystem living in total isolation, an undiscovered world thriving in darkness. I was in utter awe.

Paul and I drifted over the shag carpet of life side by side, easing into the iceberg, filming every step. This cave would make our movie special. Then, suddenly, the silence was interrupted by a strange deep rumbling that reverberated through the water. Was it the boat engine or something else? We had been swimming for about forty-five minutes, and I figured it was a good time to turn back. We had found something worthy of exploration but needed to shoot with the larger camera and biggest video lights, which we didn't have with us. I turned to Paul and extended my thumb upward, indicating it was time to retreat. We spun around and eased our way back toward a dim glow of light cascading down from above. As we neared the top, I glanced up at the Zodiac, but something had changed. Although the light still streamed down on us, it was more fractured and shadowy, and it was only as I swam up through the crack that I realized that the thin sliver of blue open water was gone. I could see only white chunks of ice in the diffuse light over our heads.

I watched Paul beneath the jumble of ice as he searched for a way out, but I still didn't realize the seriousness of our situation. Although our doorway was gone, it did not immediately occur to me that we could be trapped. Then adrenaline surged through my body in a flush of warmth, a reaction to the fear. I joined Paul in the search for a way through the chunks of ice, trying to push the large pieces aside. Finally I popped through a gap that led toward blue open water. I pried away a large slab of ice and found a place to complete my five-minute safety stop before surfacing. Still unsure about what had happened, I looked up through the water and saw Wes and Matt high-fiving and hugging. Paul and I completed our ascent and surfaced in the chilly brightness of a sunny day.

The situation had been much more frightening to our colleagues than to us. As they waited for our return, a large slab of the ice wall had peeled away and crashed into the water, narrowly missing the inflatable Zodiac. The resulting swell nearly toppled and swamped the Zodiac, sending Wes and Matt careening into the console and landing in a slosh of water in the bottom of the boat. In the heat of the sunny day, the berg was melting, and with that meltwater came instability. The remains of the ice wedge blocked the entrance to our cave. "We thought you were dead!" Wes said as we emerged from the water.

The dive debrief was sobering. Wes and Matt had been frantic with worry, knowing there was absolutely no way they could help us, that we were on our own in finding a new exit. And if we hadn't returned, there would be no search. Still, even after this scare, Paul and I didn't have a full grasp of how dangerous iceberg diving really could be. We would continue to do our work with an understanding that once we submerged, we were on our own.

A FEW DAYS later, we were back inside the iceberg. We entered the same cave and moved quickly into the tunnels to document the unusual life growing on the seafloor. This time I paid more attention to the peculiar features in the ice. Passing by layers of snowy deposits, I was acutely aware that I was dropping down through a time capsule. Every layer represented a slice of history. One layer would be white and frothy, but a few feet deeper, I would pass a band of bluish translucence. A little deeper, the ice was as clear as glass. One apparent void looked too transparent, and I thought it was a tunnel leading off into the interior, but when I arrived, it turned out to be a solid window of entirely clear ice. Once under the berg, we swam quickly toward a dim glow in the distance. With nothing to offer us a sense of scale, it seemed close and yet unreachable. We swam for ages, and the glow in the distance never seemed to get any bigger or closer.

Roach-like amphipods were abundant and busy on this dive, darting all around us. Perhaps thousands of pairs of the eight-inch-long hard-shelled creatures were raining down on us from cracks over our heads, flitting away as we made contact. It would have been perfect material for a house of horrors, but instead of feeling fear, I was fascinated with the mating pairs of bug-like animals. We let them land on the cameras' dome ports for creepy close-up shots. Then we continued to move in the direction of the light.

We were swimming for a short while when a current began to bend over the filter-feeding animals on their stalky anchors. At first I noticed we were drifting faster into the berg, and then very quickly the flow was pushing us in the direction of an illuminated doorway in the far distance. I stuffed my glove into the seafloor, throwing up a puff of silt and arresting my forward motion. My out-of-control body cartwheeled around to face Paul, and I thumbed the dive, aborting our mission. I knew this current was moving at a speed beyond our capability to swim against it, and we had to leave. But immediately, I realized we weren't going to be leaving the way we came in. The current picked up so fast that we needed to go with the flow or risk exhaustion swimming against it. Plus, having seen our doorway close over us on the previous dive, I was just as content to give in to the current and drift along, even though there was no guarantee that our path upward would be clear enough of ice to reach the surface safely. But, talking through the rebreather loop, Paul and I decided it was the best bet in the moment.

We swept past colorful creatures that seemed accustomed to the strong flow that was rushing us toward an uncertain exit. Ten minutes later we were spit out through a fracture that offered a doorway out of the berg. This time, the passage was as wide as a two-car garage, and we began a slow and careful ascent and decompression hang. I looked up and saw patches of open water and lots of boulder-sized blocks of ice bobbing around. After a ten-minute hang, I popped my head above the surface and looked around for the boat.

At first I was facing the sheer wall of the berg that towered above me out of sight. I rotated and faced a huge ice boulder bobbing three feet high out of the water. I continued to spin but saw nothing but ice in my way. There was no boat. "We have a real problem here," I said, breathing the chilly Antarctic air just as Paul popped up out of the water. I wished I could soar up above the ice as I had on my chopper ride. I reached into my dry suit's hip pocket and pulled out my emergency pouch. I unfurled an orange marker and inflated the six-foot-high mast that was equipped with a small radar reflector. It barely poked above the ice that surrounded us. We were invisible to anyone beyond these walls of ice and could freeze or be crushed before we were spotted.

I grabbed on to a raft of ice and tried to pull myself up onto the jagged surface. It was a hopeless endeavor. I was simply too heavy and the ice too slippery. I bobbed and kicked to get a better view over the lip of the frozen horizon and saw not even a glimpse of color. It was just ice. We were surrounded by a harsh white expanse. I pulled out my whistle and started blowing hard to attract attention. The surrounding ice seemed to absorb and mute my shrieking blasts. I turned to Paul. "What's next?" I asked. "This would be a horrible way to go." He seemed as scared as I was, but a little more optimistic that we would be spotted. I tried to reconstruct our underwater route in my mind and estimate where the boat might be. If I was right, it could be as much as a mile away and around the corner of the massive berg.

The minutes ticked by while I continued to kick and move to generate body heat. The day was clear and sunny, which offered the hope that the helicopter could at least begin a search for us. We had been gone for a couple of hours at this point and were sure our team would be getting concerned. Paul and I decided to inch along the edge of the ice wall in the direction of where we had begun our dive, pushing small floes out of our way, knowing that if the boat spotted us, they would still need a clear path to pick us up. Descending again was always an option in the event the ice closed in on us, but I was loath to disappear from sight. Even though there was no darkness beyond a short period of dusk in the late-night hours, and our crew could search for us for hours, we knew that we didn't have that much time. I was feeling increasingly agitated and helpless.

Finally, I heard it—the faint rumble of the boat's engine. I listened to the transmission grind into gear and knew the boat was moving. Were they searching for us? How long would they be? Then, the distinctive clanking that told me they were pulling up anchor. That was a good sign. Paul and I blew our whistles, hoping they'd hear us and that we'd soon see the *Braveheart*. The sounds of the boat energized me, giving me the strength to swim faster along the edge of ice face. Then a sliver of the stern appeared at the corner of the iceberg, about a thousand feet distant, quickly disappearing again as the boat maneuvered into place. I shouted, but there was no way they would hear me over the boat's noise.

The transmission shifted again, and then the stern teasingly turned into our view once more. This time I heard music to my ears: "Is that Jill?" I could see Wes at the rail. His eyes stayed locked on us while Matt Jolly joined him. Then *Braveheart* turned on a dime toward us, Wes and Matt running along the rail to keep us in sight. Ice floes were pushed and heaved by the bow as they worked in our direction to pluck us from the icy water one more time.

In light of our two close calls, it might seem insane that we went back inside again, but we did the very next day. Mother Nature's siren call and working in intense situations can lead you toward a strange take on risk assessment. At the moment that we were planning our next excursion, I knew the technical diving world might call us foolhardy. But your baseline of risk assessment can shift in remote places—and under the pressure of deadlines. When you survive a close call, your baseline has a tendency to shift in what is known as normalization of risk. Our brains are finely attuned to novelty and new sensations, but lose focus when things become commonplace. We stop paying attention to the less-than-novel stuff, and for expeditioners that means we inevitably slide toward complacency—and sometimes we only see that slide in hindsight. What may be regarded as almost suicidal one day could become accepted and even normal. Things that were formerly considered dangerous and risky become seen as challenges to be endured. When an entire team accepts a dangerous situation, it offers a sense of normalcy—if everyone thinks the situation is reasonable, then it probably is.

A great example is the *Challenger* space shuttle disaster. At that time, space shuttle missions had become routine. Few people watched the launches and returns in those days, in part because the risky expeditions had been repeated over and over again without incident. Space travel was no longer sexy to the public. Then, on January 28, 1986, the *Challenger* broke up over the Atlantic Ocean seventy-three seconds into its flight.

Engineers at NASA mission control had allowed complacency to slip into their routine operations after many successful undertakings. They had even predicted the failure of an O-ring seal during a cold-weather launch, but their managers chose to ignore them, arguing, "It hasn't killed us yet." They had become accustomed to the risk and altered the threshold for aborting a launch. In the post-accident analysis, the lesson was clear: safety protocols are necessary, and one must always guard against the normalization of risk.

So were we prudent to make another dive? In the context of our mission, it felt like the right thing to do. We needed the footage, and we were running low on fuel and would need to start back to New Zealand soon. After a team meeting approved our plan, Paul, Wes—who was ready to get back in the water—and I geared up for the dive. Since this would likely be our last dive in the area, we were focused on what needed to be filmed for the movie and *National Geographic* article.

Yet even the act of filmmaking can push people to take unreasonable risks. There can be enormous pressure to "bring back the goods." Few people understand how much groundwork, time, and risk go into making even a short underwater segment of a movie. We undertake years of training, invest hundreds of thousands of dollars in gear, and then break and repair most of the equipment that we use. We leave our families for months, endure seasickness, wither from weight loss, and lose night after night of sleep. We scout for weeks, get trapped in the ice, find a site, explore the site, risk our lives under a calving iceberg, and then get in again for another dive. After all the preparation and rehearsal, only half of the cinema lights end up working underwater, and other gear fails or floods. When all is said and done, we have shot what winds up being a two-minute segment that makes the edit into the final film, and still, I would not trade the experience for anything in the world.

When we were lowered into the water, there was still quite a vigorous current, but we weren't worried: as soon as we'd get into the berg, we'd be able to hide from the flow. Paul and I guided Wes to the massive fissure that penetrated the berg and descended into the darkness. Upon arriving at the seafloor, we ducked under the berg, revealing all the extraordinary life for Wes's camera. I heard him gasp in awe at the colorful red and yellow fronds that were wafting with the flow. With the current still brisk, we carefully moved inside.

Paul and I swam deeper into the passage ahead of Wes, engrossed in lighting up the crawling amphipods, spindle-legged arrow crabs, and lumpy mounds of porous sponge. But I found myself getting distracted by a sensation in my left hand. It was beginning to hurt, badly. A knife-edge of pain started in my pinky finger and was spreading out. I paused to take stock but couldn't see anything wrong.

I looked down again at the blanket of life on the seafloor, exquisite and intricate in its variety of shapes and colors. But my attention didn't last. The inside of my left glove was now filled with sloshing cold water. The gloves I was wearing were specially designed for polar diving, using a system of nested rings that close around the wrist with a latex rubber seal. Somehow, either a ring had failed or I had poked a hole in the glove. In either case, a tiny stream of water was now slowly freezing my hand, and the pain was extreme. I tried to direct my attention elsewhere, but I could not ignore the problem. I checked my wrist computer. We'd been in the water for about forty-five minutes, and if we were to turn around immediately, we'd still have another hour of exit and decompression time ahead of us before we could break the surface.

I raised my arm above my head to shift a pocket of air into the glove. The cold water slowly crept down through the wrist seal, into my sleeve, and toward my armpit. Bad idea. I lowered my hand again, and the increased pressure squeezed the air from the glove back into my suit, chilling my aching hand all over again. I folded my thumb and fingers into a fist inside my glove to keep them as warm as possible. Then I tucked the loose fingers of the glove into my fist, hoping to stem any further leaks. I wasn't sure how long I could bear the cold, but for the moment, I decided to endure the discomfort.

We were eight hundred feet horizontally inside the iceberg and I tried to focus on shooting. While Wes ran the camera, Paul and I illuminated the darkness with high-intensity video lights. I could see the endless garden below, the blue-green ice canopy above, and all around us a vast universe of dense blackness. As we moved forward, I could feel a column of water pressing at my backside with increased intensity, which alarmed me. I tried to brake my forward momentum by flaring my body and legs outward, but just as quickly I was pushed farther into the cave. I turned around and tried to swim in the direction we had come from, but discovered the current had picked up and I could make almost no headway. My left hand was searing with pain, and I was having trouble grasping the light head's handle. I turned to my partners and signaled to end the dive.

The HD camera that Wes carried weighed over eighty pounds—we sometimes called it "the beast." In the water it was only slightly negatively buoyant, but the unit was huge, and now it was like pushing an open umbrella against the wind in a storm. We tried to ease back toward the fissure, but the current was too strong, and for a full five minutes all we did was fall backward even though we were kicking as hard as possible. The flow was siphoning into the cave, and the current was increasing in strength. A realization was creeping into my mind: we were not making any progress toward a safe exit.

I pushed my compromised hand into the gelatinous bottom, watching wisps of clay mud rise into the water column like a smoking fire on a windy day. The debilitating freeze depleted my strength, while my blood vessels throbbed to keep me warm. Icy pinpricks of cold crept in around the edge of my diving mask—I was simultaneously hot and cold, kicking as hard as I could, latching on to anything that I thought might edge me closer to the soft beam of elusive daylight, closer to the light at the end of the tunnel.

We hadn't expected the currents to shift so dramatically, accelerating to the point that we were pinned beneath the iceberg, and our crewmates on *Braveheart* couldn't have imagined what we were facing—but they knew we were dangerously overdue. Despite everything that had happened in the past few days, I was still shocked by the force of the current. I suppose we had been given our share of warnings, but that revelation was no good to us now. We were trying to escape the tunnel and get to a route that would lead us up to safety, but first had to beat the forceful current before we froze to death.

The streaming water was accelerating at a frightening rate, pouring down the crevasse and funneling under the iceberg. If this kept up, we'd never get out. I checked my wrist display again. We'd been down fifty-one minutes and would face a lengthy decompression hang—if we got to the point where we could ascend. My hand now felt like a lead weight at the end of my arm, and my mind was starting to swamp with worry. How could we fight this current and get out of here? Would I lose fingers?

Over our heads, the ice on the bottom of the berg was slick, and there was nothing to grab on to except the living seafloor below us. I pushed my hand back into the seabed. I regretted destroying clusters of exotic creatures, but knew this was the only way I could move forward. Behind me, Paul and Wes were doing the same.

My senses became sharpened, more focused. I could feel my lungs working hard and my heart racing. I was kicking with everything I had and breathing so hard it occurred to me I might be at risk of carbon dioxide buildup, a particularly dangerous situation on a rebreather system. Too much CO_2 can overwhelm the equipment's ability to process it and cause a diver to pass out. It was critical that I get control of my breathing. Now was the time to put to work everything I'd learned and practiced over the years. If we couldn't manage to get out from under the iceberg, we'd be finished. Gradually our core temperature would drop, our metabolic processes would break down, and we would freeze to death.

My perception narrowed like a tunnel and I was only vaguely aware of the presence of my two teammates, who were facing their separate struggles. We were not capable of helping each other anymore. We were each in our own private hell, pulling and kicking, pulling and kicking. Every minute at these dizzying depths had consequences. If we could get to the doorway to the surface, we would have a long decompression time. Successful escape from the iceberg's cave would force us to endure even more pain and a choice: either we could remain immersed in the blistering cold water or we could surface early and risk the agonizing misery of the bends. It was ironic. Even if we could get ourselves out of the cave, we still might die, but from decompression sickness or hypothermia.

Wes yelled to Paul and me through his rebreather mouthpiece. At first I couldn't make out what he was saying. My heaving breath and my throbbing heart were deafening.

"The camera!" he yelled between breaths. "The camera! Help . . . me . . . with . . . the camera!"

The large metal camera housing was becoming impossible to manage. "You have got to be kidding me," I thought. "Fuck the camera! We're going to be lucky to get out of here alive!" In that moment, I believed only in team ethic and ensuring the safe return of all three of us. That did not include equipment, even expensive equipment.

I've lost numerous friends who put an emphasis on bringing home all their gear rather than saving their own lives. Sprinting underwater to reach their next air tank staged in the cave, they fell short while dragging a broken scooter or a flooded rebreather, when they could have returned later to fetch the gear. I was not going to risk my life to save an HD camera, no matter how important Wes felt it was. Did that make me less of a filmmaker? I was angry with him for even suggesting that the camera was as important as our lives. Paul dropped back to help Wes lug the camera. Instead of joining them, I channeled my adrenaline into finding solutions for getting back to the boat. We were still compounding bottom time at 130 feet, but we were getting closer to the point where we might rise to the surface through a large crack and find open water.

I knew what it felt like to wait. I knew what it felt like when I was powerless to change an unfolding calamity. How had I arrived at this point in my life, facing death in the most beautiful place that nobody would ever see? Had it been worth it? Had I lived as fully as possible or had I taken unnecessary risks? Would my epitaph read "brave" or "foolish"?

Having finally reached the lip of the cave, with the sun shining down in thick beams of whiteness, I felt a small victory, but now I had to figure out how to get up to the surface. With the current pressing down on us, we needed to climb the walls. They were slippery, though. I would have given my arm for an ice ax. Then I remembered the ice fish tucked in their burrows. They were now cowering witnesses to our struggle, but I decided they might have to be the first martyrs of our dive. Thinking, "Sorry to kick you out, buddy," I jammed an index finger into one of the hollows. An ice fish slithered out.

I reached up and jammed my other index finger into a hole just about a foot above my head, evicting another occupant. Pressing my body against the ice wall, I gained just enough traction to pull myself up by about a foot. I searched for the next cavity—it required concentration because the holes were nearly invisible against the white ice. Falling into a steady rhythm, finger by finger, I moved up the wall toward the sky. Following my lead, Paul and Wes were behind me. One by one, finger hold by finger hold, the fish were displaced and shot down into the void, and we climbed up.

With every foot closer to the surface, the tension was passing. I could no longer feel my hand, but at least I could recognize my surroundings as beautiful again, and I used it all to take my mind off the cold. There were pockets of dancing orange krill underneath the floating ice. Magical and varied jellyfish and two curious crabeater seals ranged around us. At one point the seals approached me from behind. When I sensed their presence and turned my head, they darted playfully to a safe zone behind some ice. I saw a gelatinous creature the size and shape of a blunted miniature football. Bands of tiny swimming cilia that ran the length of its rose-colored body were electrified in colorful sequences as it throbbed its gaping mouth, which puckered into almost human-sized pink lips. The sea life offered a welcome distraction from the pain of the cold that was shuddering through my body. When you are cold, you will do anything to distract yourself from the discomfort. Even playing with a jellyfish can help pass the time.

After a long and painful decompression, and three hours after we had gone down, we surfaced at the side of *Braveheart*. A worried look on his face, Greg Stone reached down to assist me up the boarding ladder. The air was sharp and whipped my face, and I could feel stinging cold salt on my fattened cracking lips and tongue. All I had the energy to say was "The cave tried to keep us today."

It would have been easy to give up inside the iceberg. When I was losing ground against the current, my mind was losing the will to fight. In that moment, I couldn't see a way out and I thought I would die inside the ice. But I decided to fight with everything I had until the very end. I broke the situation into the smallest possible pieces I could. When I managed to pull myself beyond a small red sponge, I marked the inch as victory. When I rested to slow my heaving lungs and regain control over my breathing, I counted it as success. All the small steps could lead me to triumph, I was learning. I just had to maintain my objective and never quit.

OVER A LATE dinner, we reviewed the list of remaining film objectives and debriefed the frightening dive. Even after experiencing the powerful current, Wes was not content with the footage he had shot and insisted we get more. Though I had yelled to him to ditch the camera, he held on to it to the very end—and he wanted more. It can take days to capture every element needed for a cohesive story, and he had done only one dive at this site. But with the sea ice beginning to freeze for the winter, any more diving would need to happen in the next twenty-four hours. So instead of sleeping, we prepared our rebreathers again and got the cameras ready to dive at a moment's notice. We carefully tracked the tides and forecast that sometime in the middle of the night, the current might slacken and leave us with enough good visibility to pull off a short dive. After everything was prepared, I finally slept, exhausted.

Two crew members were standing watch at 1:00 a.m. when *Braveheart* was again violently torn from her anchorage. The current was cleaving through the cave fissure again. Ship's Master Iain started to move our boat out of the way. Moments later, while Chef Carol watched the moon over the iceberg, it began to crumble along the fracture line. Battered by scouring wind and ocean waves, undermined from within by years of thawing and freezing, a vibration had been building within the iceberg, and the heavy currents tried to tear open its central fissure. Boom! Like a crystal wine glass destroyed by a screeching voice, it shattered.

Hearing the commotion, we ran up the narrow companionway from our berths, pulling on our parkas. In the twilight, we filmed as the iceberg that had been so unyielding came alive with motion. The menacing sounds echoed with cracks and retorts like a gun. From the center outward, the berg began to crumble in a cataclysmic fury. The larger, seemingly more stable side dissolved into small pieces, while the other half of it rose up like the prow of the foundering *Titanic*, hung there for a moment, then turned on its side and plowed down into the sea. As it hit the surface, it disintegrated, sending cascades of water and foam spraying into the sky. Chunks of ice were splintering off in all directions, and a huge rolling wave heaved up and headed for our boat. Iain had imme-diately begun moving away when the first signs of breakage occurred, so as the large wave approached, he was able to swing the vessel to take it on the bow. The *Braveheart* rose up and crashed down over the rogue wave, bobbing as we tried to catch some of the action on film.

Then, right before our eyes, and with a deathly groan, the iceberg had finally surrendered. It took about ten minutes for the torment of shattering ice and rocking waves to quiet down, leaving a minefield of crystal shards adrift as far as we could see. All eighteen of us, lined up in the wheelhouse and along the rail, were without words. I was both mesmerized and shaken by what I had witnessed. We had left the cave only a few hours earlier. Had we still been inside when the explosion took place, we would be dead.

There was no need for discussion; the warnings were clear, and we dared not push our luck any further. With the ocean's surface turning solid under the light of a full moon, we knew it was time to leave Antarctica. You must be willing to get within a hair's breadth of what you perceive as success, and no matter the invest-ment, know when it is time to go home.

SOON AFTER OUR return from Antarctica, I recognized that the time was approaching to abandon my marriage to Paul. Sharing a near-death experience had not bound us together; instead of cementing our relationship, it pushed us further apart. Our diving trips had been the most exciting part of our lives, but they weren't enough to hold us together in marriage. We couldn't sustain the excitement that we found in the field, and I couldn't fulfill the role he thought I should play as his wife. At home I felt disjointed, slip-ping further away from the person I knew I was inside, as though I was wearing a costume. I loved change; Paul needed consistency. I wanted immediacy; he needed more security.

I realized I had a vision for the rest of my life and that it did not include Paul. I had dreams that I wanted to pursue, and I knew that if I didn't get away from my marriage, I would fail to recognize myself. I needed to fully explore my own potential and still honor the good things we had experienced together.

WAITING

2003

A SEA COVERED in ice will ebb and flow, freeze and thaw, but eventually, you have to decide to swim in it or retreat. Ever more aware of boundaries set by Mother Nature and my own body, I was also discovering the limits of the human psyche.

I prefer to be the explorer rather than the person left behind waiting. Waiting is painful. Waiting is filled with distress and denial. Waiting for Paul in the Huautla resurgence for hours, while the muddy water climbed the gorge walls, had been agony for me. Waiting to heal from a close call with the bends was filled with anguish. Waiting for my teammates to catch up to me while they risked their lives for a camera inside an iceberg is almost indescribable. When you are the person waiting, each moment stretches into eternity, and the voices in your head threaten to take over your brain. Vivid worst-case scenarios emerge in your thoughts, and the longer you dwell on them, the more real they become. Waiting is always raw and emotional and hard.

In the heat of the moment you want time to speed up and come to a resolution, yet if the judgment is unfavorable, you beg for the reversal of time. Waiting is filled with "what ifs" and "if onlys." Waiting is like the mythological figure Sisyphus, who would propel a massive boulder up a hill only to see it roll back down again, over and over, consigned to an eternity of futile effort and constant frustration. In cave diving, waiting is cruelty and angst, followed by regret, culminating in life-changing moments, and sometimes Sisyphus's rock rolls back down on top of you.

In May 2003, I was attending an annual cave-diving workshop. At the end of the morning break, I plugged my laptop into the projector. I was about to give a presentation on my recent exploration in Antarctica. A disheveled-looking former cave-diving student, Julie Henson, approached me in the aisle. I saw misery furrowed in her brow. The emcee was beginning to introduce me, but my tunnel vision vaporized the crowd around us.

"How long do you wait, Jill?" Julie beseeched.

It was an odd question at an inconvenient time, but it was clear to me something was very wrong.

"How long do you wait for your husband to come home?" Julie's eyes welled with tears.

"Where's Chris?" I asked, worried that her husband was not with her at this emotional time.

"He went cave diving. Last night . . ."

A rapid-fire list of "what ifs" started streaming from her mouth. "We've been having trouble. Maybe he's with someone? Maybe he's late. He might be . . ." She went on as if the list of excuses could change the inevitable. When a cave diver does not come home to his wife at night, it is rarely an affair with another woman. It is a far more dangerous mistress, and the attraction is usually fatal.

Instead of walking to the podium I wrapped my arm around her shoulders and ushered her to the back of the room. I made eye contact with Lamar Hires, a cave-diving pioneer and owner of a popular tech diving company called Dive Rite. The emcee had stopped his introduction and the room seemed agitated. I ignored everyone but Julie, sensing that she needed my strength. I knew what it was like to wait for Paul in Huautla when he was overdue, and I sensed this situation was not going to have a happy ending.

I bent down to whisper in Lamar's ear as the crowd turned to see what was going on. "I need your help. I think we have a body recovery to do."

Like any other community, cave diving sometimes becomes mired in politics or infighting, but I've also witnessed a generosity and kindness that is unequaled. In that moment with Julie, I saw that spirit come to life. Tom Mount, a diving guru, jumped up and took my place at the podium. Robby Brown, a counselor, therapist, and cave diver, joined our small group at the back of the room, introduced himself to Julie, and then hugged her for hours. We ushered Julie into a hotel room and tried to pry the story from her. In between sobs, a picture began to take shape. Her husband had gone on a solo cave dive the previous night and had not returned. At least she knew where he was diving: a nearby cave called Cow Springs.

I got on the phone with the owner of the dive shop closest to the spring and asked her to rush to the dive site to see if Chris's truck was still there. In twenty minutes we had an answer, and it was an unwelcome one. When dawn breaks and a cave diver's truck is still at a dive site, there will be no rescue. As I had already suspected, this would be a body-recovery operation.

IN 1996 I had my first experience with a truck left at a dive site. The phone rang at Scuba West early on a Saturday morning, and I stepped away from filling tanks to answer it, hearing a familiar voice on the other end. Charlie Ellison, the local Pasco County sheriff's deputy, was a diver himself and hung out at the shop. But his voice wasn't filled with its usual cheerfulness.

"Jill, can you think of a local exploration spot where we might find a missing cave diver?" he asked me.

I rolled off a long list of obvious sites: Eagle's Nest, Diepolder, Ward's Sink, Joe and Mary's, and Arch Sink. Then I mentioned a more obscure spot, a muddy hole not too far from our store. I sent Charlie down a dusty road to a spot called Nemesis Sink.

When Charlie and his officers arrived at the small murky pond on the edge of a farm, they found what they were looking for: Legare Hole's truck, parked at the water's edge. It had probably been there since late Friday afternoon, the last time anyone had seen him. Legare's brother Stephan located a double-braided nylon dive line that led down the sandy beach and disappeared into the chocolate-colored water. They hoped Legare would be on the other end of that line.

Ninety minutes after the first call from Charlie, my phone at Scuba West rang again.

"Jill, I need you out here—fast," Charlie said. "We have the victim's car and we need an experienced diver to have a look in the water."

I'd heard rumors about the existence of an unexplored cave at Nemesis Sink, and if they were true, Legare might have entered a terminal restriction to try to explore beyond 265 feet of depth. Chances were that he got stuck and was unable to turn around in the small space. The sink had very poor visibility, so his body could be lost in the murkiness.

With Paul out of town teaching a cave-diving class, my buddy Tim and I loaded gear into my VW van and rushed down the dusty roads. The fine white limestone of the road coated the trees in a wispy whiteness, making it look like a winter wonderland—a strange juxtaposition with the May weather that was already cranking above ninety-five in the blistering sun. The filmy powder misted up through the rusted-out floorboards, coating us in a fine silt. I ran the windshield wipers dry, further blowing up the powder that drifted in the open windows.

When we arrived at Nemesis, I tried to avoid the gaze of Legare's mother. I had been trained in recovery operations but this was my first. It's the hardest skilled volunteer service a cave diver will ever partake in, and we try to compartmentalize our emotions, saving them for after the diving is done. I got in the water knowing what needed to be done while my brain fought between "I hope we find him quick" and "I hope I'm not the one to find him."

When recovering a deceased diver, there is no rush or flurry of activity. If you are a photographer like myself, sometimes your job entails photographing the body to document its position and the status of the equipment before you carefully begin the extraction. These photos can be important, because removing a body from the cave will destroy vital evidence that might provide information about the cause of an accident. Even though no one wants to look at a dead body—particularly if it's that of a friend—it is important to note everything about the moment when they took their last breath.

As Tim and I completed our last-minute surface checks, Legare's mother rushed to the water's edge. I had been trying to avoid her and the emotions that could interrupt my concentration, but now she was in front of me, tears streaming from her puffy eyes. "Just bring him back," she pleaded. I didn't know what to say, so I descended as quickly as I could. I needed to stay focused.

We followed Legare's guide line down through the blackness of the silt-laden water. The water was so stirred up I couldn't imagine what had compelled him to dive in this dark hole in the first place. We could only see inches in front of our masks. The walls and the floor weren't visible until we reached a restricted space more than 240 feet down. We kept following Legare's line, and it led into a tiny slot. At this pinch point, even more silt was suspended in the water, making it impossible to see. Instantly, I felt Legare's panic in my gut, and I knew that he was close by, or just beyond my reach. But we couldn't go any farther—we didn't have the right breathing gas to go deeper. Tim and I turned around and went back up. I had enough information to sketch out the cave for others to follow, so I'd leave the work of finding Legare's body for now, and wait for the divers that were on their way with better gases.

But the wait would not end for Legare's family. They agonized by the water under a small pop-up shelter for eight days before relenting. The visibility underwater wasn't clearing, and after more than 120 excursions into the depths of Nemesis Sink, the search was called off and they went home without their son, brother, and friend.

Nine months later, the wait finally ended for Legare's family. His friend Larry Green returned to the site to see if he could find the body. The chilly winter had improved the water clarity, and Larry found what was left of Legare's body wedged into a tiny slot in the ceiling of the restriction, well off his guide line. He must have lost contact with his line and then scratched and clawed his way into his crypt while trying to find the surface. He died in a quest to set foot into the unknown.

While I comforted Julie Henson in a hotel room, Lamar Hires pulled her husband, Chris, from Cow Spring. He was found floating in the cave, his body pointing toward the exit, with considerable gas remaining in his tanks. He had probably fallen asleep and drowned as a result of a sleep disorder that should have ruled him out for diving, especially alone.

WE RARELY GET a sense of a diver's last moments before they drown, but in January 2007, I had a very intimate look. Ron Simmons was a fixture in our group of friends. He was a prolific dry caver and accomplished solo cave-diving explorer, frequently spending weeks and even months staying at what we called the Skiles Compound, near Ginnie Springs. Wes and Terri Skiles had offered a home to many of us over the years. After my marriage to Paul had finally ended, I left Hudson, Florida, and everything with it behind, taking with me just our old travel trailer. I lived frugally, plugging into an outlet beside Wes's small office building while we shot and edited movies on projects that took us around the world. The Skiles family and transient friends cooked as a group and spent evenings playing darts or sharing travel stories.

Ron had explored caves in remote corners of Laos, invented ascending gear that we used for climbing, made early tank harnesses and backplates out of stolen stop signs, and drew the most accurate and beautiful cave maps I had ever seen. He scribed the details of his dives in a series of weatherproof logbooks, and those hardbound volumes held his jealously guarded history of exploration. Each night I watched him scribble and then squirrel away his books in a hiding place in his guest room. Nobody was ever going to invade the garage suite he slept in, but Ron was a little paranoid that one of us might steal his cave.

I had recently moved my travel trailer off the compound and onto my newly purchased property next door when I answered an early-morning phone call from Wes.

"Ron didn't come home last night," Wes reported, his voice cracking.

I knew what this meant. "Do you know where he went?" I asked. Ron rarely divulged the precise location of his planned dives, so this question felt a little futile.

"I dug through his room for his logbooks and got some solid leads. But maybe he's met a girl or something?" I knew that was extremely unlikely. Ron had a laser focus on his diving exploration and thought of little else on his forays to cave country.

"I'm coming right over," I said. I filled a thermos with coffee, grabbed my hiking boots, and jogged a few hundred yards through the woods separating our properties. Wes was already sitting in the driver's seat of his big white van, waiting for me.

The drive was eerie and awkward. We talked about possible scenarios for Ron's absence that neither of us believed. Was it denial or simply an attempt to protect each other from the inevitable truth that our friend was dead? We drove an hour along the cave divers' highway, U.S. 27, and finally pulled down a weedy road marked with a sign stating that the Suwannee River Water Management District owned the property. Wes had offered this exploration lead to Ron years earlier, and he had been methodically mapping it ever since. The cave was too small for more than a single diver. Although the passages were thousands of feet long, they could be negotiated only by the boldest explorers with the most minimal equipment. In the low space—not much taller than the girth of a diver's helmet—shoulder blades snagged the ceiling and your belly scraped the floor. Tanks had to be carried sidemount style, aligned with your body beneath each armpit, in order to squeeze through.

The morning light streamed through long, gray runners of Spanish moss and live oak trees that were sporting their silvery winter barrenness. We jostled and bumped the van down the road to a small clearing by the scanty spring in the woods. There sat Ron's gold pickup truck, his shoes carefully placed and awaiting his return. Wes lowered his head onto the steering wheel and wept. I placed my hand on his shoulder, but there was no comfort to offer, just company. We knew Ron was dead.

We silently stepped out of the van, then held each other and cried under the sunbeams that illuminated the dewy forest like a misty cathedral. It was peaceful, and we wanted to recover his body without fanfare and with respect. But I knew that neither of us could imagine getting in the water to recover Ron's body.

"We can't do this," I said. "We just can't do this one."

"I understand," Wes replied. This was a task too personal for either of us to take on. It required the best skill and experience; another diver from a group of Wes's friends we called the Mole Tribe.

I took Wes's cell phone and called our close friend Mark Long, a veteran explorer who was no stranger to recoveries. It was a huge request, asking him to take on the grim task of bringing out the body of another good friend. We knew it would be risky, and that Mark could just as easily run into problems that would get him killed. We knew the emotional scars of seeing Ron's dead body would be with Mark forever. We knew it was one of the biggest sacrifices we could ask from him.

"Mark, Ron is dead. We need your help." There was no point in dancing around the issue. It was a time to be direct and clear.

Mark was a strong and unflappable diver, but he always looked like he carried the weight of the world in his eyes. He sighed through the phone and asked if we knew what had happened. We always want to know what happened. It comes from the need to find order, to place blame or say to ourselves, "That wouldn't happen to me." Recovery divers have to prepare their mind as much as their gear. They will be looking death in the face.

While we waited for Mark and the police, we reflected on Ron's life. He was a quiet man and a compassionate Buddhist. We had talked about death and he did not fear it. He had told me that dying would be a release from the limitations and suffering of his physical body. I later read that the soul of a Buddhist is said to remain close to their body for several days after death, until the being experiences a flash of enlightenment. As I stood quietly in the tranquil woods, waiting for the chaos of a body-recovery oper-ation, I felt Ron's presence, right beside Wes and me under the oaks. I would try to remember this feeling and hide the images that would soon follow.

The next hours were a blur. Friends sheltered us with a pop-up tent and brought food. Mark set up his cave-diving gear beside the spring and disappeared into the small hole that was not much bigger than a kitchen table. Then we waited for his safe return.

An hour passed and then someone yelled, "He's back!" and I saw Mark's bubbles weeping out of the cave. After another ten minutes—though it felt like much longer—Mark's head slowly broke the surface. "Are you ready? I'm going to need some help here."

Wes and I knelt on the muddy bank next to him. Some others took the cue and stepped to the water's edge to help Wes and me pull Ron's rigid body, heavy with equipment, onto the bank.

"Try not to look at his face," Wes advised me. But I wanted to see Ron's eyes one last time. I wondered if they would tell me about his last moments, if they would show fear or resignation. I expected to see a ferocity that would prove that he fought to the bitter end. But when I looked at him, all I saw was an empty grotesque vessel. His eyes were swelling inside his blue-rimmed silicone dive mask.

Wes and I inventoried his equipment while someone else took notes, carefully observing the presence and functionality of each piece of gear, hoping we would find some reason for his death. Drowning is disfiguring. I was looking at a corpse that bore almost no resemblance to the man I had known. The challenge for those involved in the recovery is to try to recall the living face and not dwell on the image of what no longer looks like a friend.

Wes and I placed a towel over Ron's face. Our friend was gone but every step of the process was conducted with dignity and care. It is hard to hold your dead friend, but I also felt a sense of honor and responsibility to take care of him, even if he was just an empty vessel now.

A piece of us went with Ron that day.

Julie Henson's question about waiting was left hanging in anguish. "How long do you wait for a cave diver who has not come home from their dive?"

Forever.

7 R

2006

WHEN I DESCRIBE the act of cave diving to most people, they think I have a death wish. Why would anyone want to spend all their spare income to enter a world of complete blackness where a single mistake could leave you dead?

But I am not alone in my seemingly bizarre and boundless motivation to experience pleasure and novelty. It turns out that my genetics may be driving me to seek adventure and partake in what is seen as risky behavior. I am attracted to other people with a similar genetic makeup, the one that includes the *DRD4-7R* allele. That single gene plays a crucial part in the genome of explorers and is perhaps the very foundation of human existence. Found in 20 percent of us, it is tied to the brain's ability to regulate dopamine, a chemical that is connected to our understanding of risk and reward. Those of us with the *7R* gene don't get as much pleasure from everyday stimuli. Foods are not spicy enough. We seek variety, flavor, and zest. We tend to be more curious than everyone else, looking for newness in everything we do. We're wanderers who are never satisfied with the status quo. We explore new ideas, travel to new places, experiment with drugs and sex, new relationships, novel foods, and anything that comes with pleasant sensations and arousal. Researchers say that *7R* carriers are also more likely to suffer from ADHD and addictions, seeking an ever-increasing rush of dopamine to satisfy our bodies and minds. And if we survive our exploits in adolescence and make it to adulthood, we might even live longer than the rest of the population. It turns out that our desire to explore keeps us active enough to improve our overall fitness and longevity.

For me, the 7R gene expresses itself in my love of learning. I despise the status quo and seek change and improvement in myself and in the world. I often dive alone at my own speed, spending hours in a cavern with my camera, awaiting the perfect beam of sunlight to stream into the darkness. But the 7R gene doesn't make us reckless. We embark willingly on these new experiences, recognizing the risks. Perhaps we're even attracted to the fear.

My friend, psychologist, and fellow cave diver Bill Oigarden looks at underwater explorers from the 7R angle. His doctoral thesis examined personality traits in cave divers and how they were expressed in family relationships. He wanted to know what personality features were necessary to be able to succeed in an extreme environment. I would fill out the lengthy questionnaires he passed around at conferences over the years and wonder what he might learn—about me and others like me. When he first told me that I was a novelty and sensation seeker, I took exception to the label. It felt judgmental. I didn't want to be perceived as an adrenaline junkie. Still, in my heart, I knew he was onto something. I need learning curves, stimulation, and challenges in my life. And I wasn't alone either. Based on Bill's research, it turns out that a lot of my fellow cave divers—but not all explorers—have similar qualities. It takes a 7R to understand a 7R. We thrive in each other's company as long as the novelty is reciprocated. That's why meeting my second husband, Robert McClellan, was such a gift. He was definitely a 7R, my first 7R romantic partner.

In our first ten years together, Robert and I embarked on many crazy projects. We built a yurt, rode our bikes 4400 miles across Canada, and bought and sold four different travel trailers and fourteen bikes. We reared chickens, learned how to become bee-keepers, made soap and leather belts, built a geodesic greenhouse, farmed our food, lived on a houseboat, made documentary films, and hosted backyard concerts. Robert was a nurse at a veterans' hospital when I met him, and he used that career to travel all over the U.S. as a contract nurse. But that was not his first gig. He has worked for Electric Factory Concerts and 20th Century Fox, had a career in the navy as a photographer and instructor, and led a medical battalion in the National Guard. He has been a bartender, prison nurse, concert promoter, hippy boutique and ice cream truck owner, truck driver, stagehand, radio talk show host, combat photographer, cardiac technologist, e-commerce guru, author, music producer, soundman, alcoholic, drug dealer, and sobriety coach. Yes, he's confirmation that people who suffer from addiction issues are often 7Rs too.

When I first encountered Robert's profile on Yahoo! Personals in 2006, it clearly stated that he was a recovering alcoholic and addict with almost twenty years of sobriety. What might have scared some people off was at least slightly intriguing to me. I knew little about addiction (negative ones, at least), but I was struck by the frank honesty and guts it took to post such a clear statement about himself. I had proudly written on my own profile that I was "comfortable in my own skin." After leaving Paul, I finally felt free to be me. I would arrive for a date with all the bumps, bruises, and dirt under my nails that were my hallmark.

After I spent two years living alone in a travel trailer and slowly building a house in the North Florida woods, my girlfriends had convinced me that it was time to start dating again. My divorce was long finalized, and I had just completed a nearly year-long job on a Hollywood movie, *The Cave*, as the underwater unit coordinator. It was a fun job building rebreathers, training actors and crew, organizing stunts, consulting on the script, and pulling off complicated scenes either as the technical director or as a stunt diver. For six months I lived in the well-appointed Hilton hotel in Bucharest while we shot on a nearby studio lot. The luxurious lodging was great, but I barely had time to rest my head on the crisp Egyptian cotton sheets. The days on the set were long and my rare days off were spent exploring incredible vistas in the Carpathian Mountains, the Danube Delta, and eastern Europe's cosmopolitan cities.

When I checked out of the Hilton, the desk agent presented me with a bill for "incidentals." First I gasped at the 320-euro supplement, and then I asked what sort of incidentals this pertained to. The desk agent scrolled through her computer screen and checked the tallies on several months of small charges, then loudly declared that I owed the hotel nearly $400 for condoms that had been placed in my room by the housekeepers. In earshot of a dozen people who were now looking at their shoes while snickering, I responded, "Trust me, I can assure you that there has been no reason for anyone to use a condom in my room in the past six months!" I refused to pay the charges.

So when my close friends encouraged me to try online dating, I agreed it might be time. For anyone who has been through the experience, you will know that you spend hours filling out questionnaires supposedly designed to whittle down your scope of interest. They want to know how you spend your spare time. They want to define a geographical radius for how far you will travel. They want to know about things that are deal breakers. So when you spend an evening filling out the endless surveys, you get a sense that the result will be a reasonably close match. At times, the process can be quite gut-wrenching. When you are asked to pick an age range, you reflect on whether you are a cradle robber or have a daddy complex. Every question leads to extensive self-examination, with the awareness that if you are not honest, then you won't find a suitable match.

By the time I completed four hours of introspection, I was excited when the machine finally started crunching the numbers.

"Almost there! We're searching for your matches." I'll admit my heart was racing. What would the algorithms reveal? Would he be a handsomely weathered Indiana Jones? The spinning clock continued to sift through possible matches. It seemed to be taking an eternity, and every moment that passed made me feel ever more naked and weak. Finally, a chime awakened me to the moment that would guide my next chapter in life. My 7R kicked in. Who would be my digital prince?

"We're sorry—we are unable to find the right type of people for you."

"What!?" I yelled in my terminally lonely bedroom. "How is that possible?"

I deflated into a slouch of humiliation. I deleted my eHarmony profile, posted a three-minute effort on Yahoo! Personals, and headed off to the Florida Everglades to work on a documentary film. Working with Wes always cheered me up. Although we worked hard, we had fun. More importantly, he always encouraged me to be my best authentic self.

Wes and I had finally transcended a recent run of work on cheesy reality TV show gigs and were partnering on a PBS documentary series called *Water's Journey*. I was writing and producing the programs, and Wes was directing and shooting. With a meager budget, we also filled the roles of on-camera personalities and covered all the jobs from graphic design and PR to editing and effects. I felt a great sense of purpose in our work together. The first two *Water's Journey* films had won international acclaim and were making a big impact in the education system. As a result, our public profile as water conservation advocates was growing.

The premise of the series was to follow a drop of water through the environment wherever it would lead us. Although we used cave diving as a hook to draw in the viewers, the series was intended to teach people about water conservation. Sometimes we used a radio-location team to track our path, much like we did at Wakulla

Springs. With a topside tracking team in full bush gear, Wes and I swam through the caves that passed beneath some pretty unusual locations. We traversed under forested landscapes and swamps, wove our way under urban jungles, and swam below a bowling alley and a golf course. In the first episode, my diving partner, Tom Morris, and I swam through a cave under a Sonny's BBQ restaurant in Alachua, Florida. As we did that, Wes and our Wakulla Springs radio tracker, Brian Pease, burst through the door of the restaurant yelling, "Cave survey team coming through!" They had just climbed up a steep embankment from a swampy ravine, hacking branches out of their path until they got to the restaurant's parking lot. They wound among the booths filled with stunned diners who weren't sure why a man was carrying a machete and little orange marking flags. While the camera was rolling, Wes turned to a waitress, nodded, and planted a flag in the potato salad. Beyond humorous, the scene showed that drinking water was running beneath us wherever we were.

Our *Water's Journey* installment on the Florida Everglades was a two-hour episode that examined water issues like stormwater runoff, fertilizer pollution, and overuse from the headwaters near Orlando's Disney World all the way through the South Florida ecosystems that make up the Everglades watershed. We shed light on horrifying green algae that haunt the Lake Okeechobee watershed to this day. We hiked, kayaked, sailed, swam, dived, and used airboats to make the trip, encountering what seemed like half the alligator population of the state in the process.

While I trekked through swamps spying Florida panthers and myriad snakes, my new Yahoo dating buddy, Robert, and I built our new relationship via email. As odd as it might seem, my absence from civilization offered us the pleasure of months of letter writing to get to know each other. In that way, our budding romance was a lot like an old-fashioned courtship. Robert was romantic, writing me thoughtful notes and even a funny song about making movies. He had a great sense of humor and the patience of a saint, saying that he would wait as long as needed for us to have the chance to meet face-to-face. As we exchanged pictures, his arresting but gentle eyes seemed to reach out from his tough, muscular exterior.

One of my favorite days on the project included filming in a remote region of the Fakahatchee Strand. With a state park guide leading our efforts, we waded through thickets of pond apple, cabbage palm, and cypress, searching for rare ghost orchids. Like other orchids, the ghost variety is attractive to collectors because of its magnificent beauty and unabashed sexuality. The pale voluptuous bloom opens phantom-like in the night, its tendrils unfurling into thin strands. It resembles the silhouette of a leaping frog caught in mid-flight.

Our guide, Mike Owen, knew of seven closely guarded specimens in the speckled light of the humid woody canopy but rarely shared their location. Sought after by collectors, these plants did not make themselves easy to find. The orchid's tangled web of tubular roots clings to the bark of swampy vegetation, but can thrive only in the presence of a rare and unique fungus. Filtered sunlight activates the air root growth year-round, but the plant will only bloom in a nearly miraculous combination of events. For less than three weeks during peak mosquito season, if all goes according to plan, the giant sphinx moth will hover admiringly at the beautiful bloom, then unfurl its ten-inch-long proboscis and slap its tongue deep into the spur of the orchid. No other insect can reach the sweet and tempting nectar, yet somehow in the Fakahatchee, the hummingbird-sized giant sphinx had managed this year to pollinate seven blooms, each known by our guide with a simple reference number. There are years when nobody reports finding a single specimen and other times when known plants are poached by nefarious agents of illegal collectors. On this day, after a recent hurricane, we found one plant, entwined on the trunk of a tree hanging precariously just above the water. We couldn't possibly peel away the plant but decided instead to prop up the wind-damaged trunk and hope that the symbiotic tree, fungus, and moth would all work together to save this endangered plant. It was unlikely that the giant sphinx moth would find this fascinating bloom so close to the water's surface, but we gave it our best.

During this memorable trip, I began to see my collaborative partner and backwoods brother Wes as the fragile and endangered moth. My giant of a friend was capable of creating the most unusual and unique symbiosis, bringing together remarkable and talented people in a way few could accomplish. He oozed creative genius and a sense of adventure that ushered our team to notable places around the world. He had a unique way of enabling all his friends to find their best missions in life. He stimulated in me an understanding of my role in water advocacy. Through his example, he gave me the confidence to fully pursue my creative career. More importantly, he reminded me that someone worthy of my love would be supportive of my dreams.

But I was starting to notice a decay in his otherwise healthy physique. His face was losing its joy, sometimes replaced by pain, confusion, and even anger. Back injuries, migraines, and a chronic bowel ailment from excursions to Haiti caused him a great deal of suffering. He was only a few years older than me but was aging rapidly. His physical torments would leave him unexpectedly sidelined, in bed and unable to interact with anyone. It was hard to watch and tough to manage under the deadlines of production and editing. I was never quite sure if a filming day would be postponed, leaving me with a crew of people lingering around his big white production van awaiting a possible departure. It was excruciating to watch my dearest friend slipping away, but even worse to see it happening in view of his colleagues.

On a day of aerial filming in Central Florida, I started to realize Wes's issues went far beyond exhaustion and pain. We were sitting on camp mats on the vaporizing airport tarmac assembling the Sony F900 camera onto a gyroscopic mount that would enable us to control the camera from the helicopter cockpit. It was so hot that I flinched whenever my exposed skin touched the asphalt. Sitting cross-legged, passing camera parts to Wes, I noticed that he was not fully engaged. His eyes were drooping and he was slurring his words. Helicopter time was extremely expensive, so our usual drill was to assemble the parts as quickly as possible, back-focus the camera, and then take off. Wes would sit in the copilot's seat operating the pan and tilt motors, while I melted in the back seat with a large piece of black Duvetyne fabric draped over my head so I could see the camera monitor in the bright sun. Sweating under the cloth, I would direct the sequences, spotting bug splats that would have to be removed during frequent landings to clean the lens.

On the tarmac Wes was distracted and confused, seemingly unable to put the camera parts together to his satisfaction. A state water management official was briefing us on the various dams and canals that we would film from the sky, but Wes was slow and distant. He answered his ringing cell phone and spoke in a protracted drawl that was thick and stagnant. As I watched him speak, the screwdriver in his hand plinked to the ground, and his eyes rolled back. He slumped forward and began to snore, the cell phone still held to his ear. Then he awakened and looked up, smiling with childlike innocence. But I was worried. "Let's get you a cola," I said. This was not the first time I had seen Wes fall asleep at inappropriate times. Something very serious was happening.

The entire week had been a struggle to get the last remaining footage that we needed for the film. Wes was passing out at the dinner table and was no longer able to drive his van. Although he still rallied to shoot stunning footage, he would then slip away and lie down anywhere he could find a flat spot to rest. I had to cancel diving shoots and had only one option ahead: to halt the hemorrhage of production funds. I canceled the remaining filming days.

TOP: Paul leaves Ice Island Cave #4 in Antarctica after our first historic dive inside the iceberg.

BOTTOM: Paul weaves through icy tunnels inside an Antarctic iceberg.

TOP: Running low on fuel, the crew of the *Braveheart* ties up to an iceberg called Patience Camp II to begin diving and research activities.

BOTTOM: The *Braveheart* pushing south through thick pack ice in the Ross Sea, trying to reach the B-15 berg and avoid getting trapped.

TOP: Walking with penguins at Cape Hallett in Antarctica. *Photo by Paul Heinerth.*

BOTTOM: Robert McClellan and I promised guests "jerk chicken with vows" on our wedding invitations. The backyard wedding, on April 21, 2007, was the best day of my life.

TOP: While filming the PBS TV series *Water's Journey* in Daytona, Florida, Wes Skiles and I went straight into the eye of a hurricane.

BOTTOM: This was the last photoshoot I had with Wes, one month before his death, and I'm grateful for it.

OPPOSITE: Diving brings many wonders into my life, like the chance to explore this WWII shipwreck that was sunk by a German U-Boat in Bell Island, Newfoundland.

TOP: Whether it be stingrays, sea lions or polar bears, having the chance to document rare, endangered and charismatic marine life is one of the greatest privileges of my career.

BOTTOM: My favorite cave on Earth is Dan's Cave on Abaco Island, Bahamas. I captured my cave diving brother Dr. Kenny Broad exploring this stunning crystal palace.

OPPOSITE: Another corner of Dan's Cave. Here, I was joined by both Kenny Broad and Tom Morris on a *National Geographic* exploration, survey and outreach project.

TOP: A little glimpse into the future. The Sunfish robotic mapper, developed by Dr. Bill Stone's Stone Aerospace, made the first completely autonomous, 3-D map of a cave in the fall of 2017. Soon, it will be able to complete cave diving missions that are far beyond my reach.

BOTTOM: A favorite self-portrait. Here I am descending, as though through fire, into Devil's Ear Spring in Florida. The water is made vibrant by tannins from the Santa Fe River.

Wes had become a danger to himself, and I worried about how to answer inquiries from our state government sponsors. "What's wrong with Wes?" one asked. I had no answer.

After Wes fell asleep in his meal one evening, I nervously entered his room in the seedy strip motel we were staying at. I knew he would not be happy about it, but I felt I needed to protect him and his reputation. "Wes," I said, even though he appeared to be in a stupor, "I've told the crew that we'll head home in the morning. You're not well enough to keep this on the rails."

He was suddenly wide-awake and I saw the fury turn his face crimson. "Just let me make my fucking movie!" he screamed. He came right at me, trapping me against the wall while he spat a tirade of savage commentary I was sure he would regret. It was absolutely terrifying. I had only once before seen him out of control, and not toward me. I was no longer looking at the face of my closest friend; I was gazing into the terror of addiction. Whether it was the opioid pain pills he was taking for his many ailments or something else, this was not the man I knew. This was the screaming helplessness of pain and the desperate attempt to make it stop. I ran from his room and locked myself in my own. I slid down to the floor with my back braced against the door and sobbed.

It pains me to recount that time in his life when I unwillingly became one of many enablers to his addiction. I made excuses for his behavior for a long time before I realized he was hooked on pills that had originally been prescribed by a doctor. At times, he asked for help and surrendered his white-capped orange bottles to me and another crewmember to dispense on their prescribed schedule. But he was only trying to appease me temporarily so we could complete our projects. There was always a second supply stashed elsewhere, and I would discover single tablets hidden in the pockets of our camera bags.

It was my new beau and recovering addict Robert who helped me understand that it was the drugs talking, and not my fallen friend. "Jill, we addicts can create a tornado around ourselves, and you have to steer clear of the vortex or it will suck you in too," he told me. "Somewhere inside, there is still a Wes that you know and love, but right now you can't help him. He will have to crash hard if there is any hope of recovery. You must get away, and keep a healthy distance."

The Wes Skiles I knew and loved from our years of work together had helped create the rules for safe cave diving, spent endless energy protecting Florida's springs, captured some of the most iconic imagery in cave diving, and was a good friend and confidant. We had danced a creative symphony together, traveled halfway around the world, and fostered a mutual respect of encouragement. But I knew our time together was waning, and we would each need to embark on our own new creative directions. After the Everglades shoot, we still dabbled in projects together, but only when I thought he was gaining ground against his demons. We still brewed up new ideas and proposals, but failed to follow through on most of them, knowing that in some ways we had outgrown each other or that I simply knew too much. Wes seemed more comfortable moving away from old friends to nurture new, naive relationships that satisfied his 7R temperament. Perhaps my 7R gene helped me drift away, too, into the new and exciting projects I had created on my own. In some ways I was incredibly sad that we were no longer able to produce art together, nurturing our symbiotic blooms of creativity, but I also sensed a darkness lurking.

MY LAST DIVE with Wes came in 2010, while I was writing a technical diving manual. Wes was a true pioneer in cave diving, and I wanted to interview him for the book. We still lived on opposite sides of a fencerow, so I walked through the ten acres of piney woods to speak with him in the office we used to share. The trail had grown over with my less frequent visits, but the green forest was fresh in the morning dew. When I got there, things were a bit awkward between us—it had been over a year since I last sat in our office—but Wes soon brightened up. He regaled me with the details from his early cave diving, and I took notes. Then we decided to go for a dive so I could photograph him in the cavern zone of Devil's Eye Spring. The air smelled the way it only does in the carefree mellowness of early summer: smoky campgrounds, sizzling barbecues, crisp air. The verdant canopy of cypress cast speckled shadows on the beams of sunlight that lit up the rippled sand of the spring run. The water was so clear that the humid summer air seemed hazier than the clarity beneath the surface. Although we both lived across the road from the spring, we could have dived it another thousand times and never grown tired of its beauty.

Wes looked angelic in the water. He moved with a natural grace that proved that water was his true element. Even in 150 pounds of gear and hefting a 30-pound camera housing, he looked elegant. Submersion always returned his troubled brow to a simple youthfulness, and his crow's feet pinched in response to his underwater smile. It felt good to be in the water with a man I considered a brother. We had shared so many great moments of travel and exploration. We had held one another when Ron Simmons and Henry Kendall and so many other friends had died. We had made important movies together, talked to schoolchildren about water issues, lobbied government officials, and supported each other in trying times. In that moment, in that clear spring water, all the torments washed away.

A month after that dive, my phone rang, and I picked it up to hear the frantic voice of a close friend and former assistant to Wes: "I think Wes is dead! We're not sure yet, but I think he's dead." Feeling suffocating panic, I quickly ended the conversation so I could call his cell phone. "Wes, pick up. Pick up! Please call me right away!" I called three times, each time leaving another message with greater disbelief. I wonder now if the phone was ringing on the boat deck while his oldest friend, Scott, bloodied his knees while giving Wes CPR. I could imagine other friends were calling too, making a last attempt to reach a friend who had been gradually leaving us all for years.

Wes began his last day filming two scientists who were documenting the feeding behaviors of goliath grouper off the east coast of Florida. He had donned an unfamiliar rebreather unit, borrowed from an intern, and casually dropped down on a dive that was well within his capability. But on this day, he had put the filming objectives ahead of his own safety, making poor choices that would kill him. He signaled his buddies that he was going to swim back to the boat for more tape or batteries and never returned. An hour later, the two scientists found his body on the reef and made all efforts to save him.

It was perhaps the most predictable, but equally one of the greatest, shocks of my life, one from which I could not quell the emotional tsunami. Wes had been the most significant catalyst in our 7R diving group. So many of us had experienced our most exciting, most frightening, wild, authentic life events with him. He was a spark that set our creativity afire, that ignited the child in all of us. He was a teacher from a family of educators. He mentored and nurtured us to do whatever made us whole. We listened as he sang at the top of his lungs, hollered in wild abandon as he held a lit firework in the air, and watched him drag many branches to our evening campfires. We had shared meals in his kitchen, loved his family, and learned to live fully in the presence of his example.

But in Wes I still see the giant sphinx moth flying too close to the elusive orchid hanging on to a subsiding branch. I watched a sputtering sphinx moth wet and damage his wings, crashing hard into the transparent surface, to surrender broken and in complete exhaustion. The orchid, unpollinated, its roots now drowning, submerged into the depths of the Everglades, never to reveal its beauty again.

We said goodbye to our fifty-two-year-old friend at the largest memorial gathering I have ever attended. From around the world, people gathered at his last campfire, a Viking pyre set ablaze on a raft on the surface of Ginnie Spring. I remember watching his beautiful daughter floating in the water, silhouetted by the flames as sparks soared through the cypress branches and into the starry night sky.

My mentor and friend was gone; a rare flower's last bloom.

CORK IN THE BOTTLE

2011

NOT LONG AFTER Wes died, I accelerated my work on my independent documentary about water literacy. It was an homage to my mentor.

But while I was able to celebrate the good works in Wes's life, I was having trouble accepting the destructive forces that led to his death. He made a cascade of choices that seemed like a slow-motion suicide. I wanted to believe that I was different, that the same wouldn't happen to me. I was a true survivor, wasn't I? I learned from my diving mistakes and recognized there would never be a superhero to sweep in to save me if something went wrong. I knew that my survival would ultimately depend on physical skill mixed with a strong will to live. As I often told my cave-diving students, "Survival doesn't have to be pretty, just effective." In the face of danger, take a deep breath and get to the work of survival. Finish the job and get home safe, in any way you can.

My dear friend the cave explorer Woody Jasper was a great example for me. In May 1990, he was picnicking at Otter Springs, in North Florida, when somebody yelled for help. Despite their instructor's warning, three newly minted open-water divers had swum into a cave and become lost. Always prepared to dive, Jasper abandoned his lunch and ran to his pickup, where his scuba tanks lay waiting in the rusted truck bed. He quickly geared up and then plunged into the blackness of the silty cavern. Within minutes he had located two unconscious divers, legs dangling below an air pocket in the ceiling. As he rose, he noted the closed eyes inside the divers' masks. He dragged the first body out of the cave to the sloping shore, where a crowd of bystanders hauled out the victim and miraculously revived him. Quickly returning to the pocket in the ceiling, Jasper found that the second man had awakened. He brought him out of the cave and resumed the search. The last diver was harder to find, and when Woody finally did, it was too late. Jasper saved two of three divers and landed himself on an episode of the reality TV series *Rescue 911*. The cave was later closed to diving and opened only on rare occasions to scientific and safety teams. I had one of those rare permits, and Woody would once again be called to this cave for an emergency. This time it was for me.

In January 2011, I had arranged to film Ruth, a young marine biologist collecting samples of strange orange bacteria that were coating the walls of the cave. As we prepped our gear in the pouring rain, I learned that my trusty pink Omega scuba regulator from 1988 was older than Ruth was.

Our first two dives went without incident, although I was transfixed by a dead turtle stuck in a crack in the ceiling, close to the entrance of the cave. "You were so close, my little friend, you almost made it," I thought. I zoomed in to film his tiny webbed foot, which was waving in the flow of the rushing water. I thought the footage might be powerful in my conservation documentary.

I decided to end our second dive at the first significant pinch point in the cave. The cave was small and silty, tall enough only to push through while grinding our backs on the ceiling. It was clear that passing that narrow restrictive area was going to be a challenge. Ruth was eager to continue, but I was a little apprehensive about diving with a new partner and wanted to top up our tanks for the task ahead.

On our third dive that day, we entered the fissure that dropped down to the low opening of the cave. Chunks of algae sloughed off the walls, engulfing the motionless turtle as I passed him for the fifth time that day, and the tunnels were slightly silted from our previous incursions. The disturbed orange bacterial deposits were now obscuring some of our visibility. Finally, we reached the turn point from our earlier dive and sized up the low restriction. I eased myself through the fourteen-inch-high space, trying to disturb as little of the silt as possible. A little at a time, I rotated my ankles in tiny frog kicks. Ruth followed, carefully slipping in behind me. The space was wide but extremely low, with only one narrow route forward. To our left and right, the ceiling squeezed down even closer. The ocher-stained guide line that Woody Jasper had laid decades earlier was positioned with intricate precision to keep our bodies located in the largest area possible. We continued swimming through multiple restrictions, finning as little as possible in an attempt to preserve the clarity of the water. But it was clear that the cave was too small and tricky, so, deciding we had gone far enough, I rotated my body like a helicopter to face Ruth. I shone my light on my upturned thumb to signal the end of the dive. She returned the gesture, rotated her body to point out of the cave, and then, with a big fin kick, obliterated the remaining visibility. I immediately and gently grasped the guide line between my thumb and index finger to follow it out of the darkness. I was accustomed to diving in poor visibility; this was no big deal. As we say in cave diving, "Silt happens," and some caves are impossibly difficult to keep clear—that's what the guide line is for.

But my calm demeanor quickly turned to alarm as I realized the line in my hand was getting tauter, stretching to its very limit. Now it was jerking sideways, tightening as Ruth struggled in front of me, farther along the tunnel, scraping against the low ceiling. Knowing she had snagged the line, I lunged forward and grabbed her ankle to pull her backward, but I startled her, and my effort was met with more frantic finning. She was panicking, her leg kicks stirring up even more silt. I tried to pull her back into the larger part of the cave, but she was fighting me. I don't know what was going through her mind in that moment, but she was losing control. Moments before, Ruth had been someone I trusted, but with her sudden panic, she was now the cork in the bottle containing my life. She was intent on getting to the exit, unaware that she was entangled in our line, the only thing that connected us to safety.

I launched myself forward and alongside her and grabbed her hand. I squeezed it, trying to convey a sense of calm while attempting to pull her back and ease the tension that gripped her mind as tightly as the braided nylon line in my hand. She continued to fight my efforts—and then it was too late. I felt the snap of our lifeline. Now we were stuck, in a dark, watery cave with a broken guide line. The visibility was near zero. There was no way to know how far Ruth had pulled us laterally off the main route and into the dead end of the terminally sloping ceiling. I held her leg with one hand and the bitter end of a broken guide line in the other.

At moments like this, your body screams for more air. I heard Ruth's respirations ramping up, and she yelled to me in the blackness as I tried to release her from a web of old buried guide line lying beneath the silty bottom—long discarded remnants from early explorers. Despite being just inches apart, we couldn't see each other, there was too much silt. My fingertips became my eyes, trying to build a picture of our unfolding emergency. I felt the tangle of line and the oozing clay floor and tried to send relaxed vibes to her while holding her trembling hand. Then I felt a shift in her. Her firm hand became weak. "Don't you give up! Don't you give up!" I yelled.

I worked quickly in the darkness to release the last bit of nylon line, and we moved forward. I unclipped a small reel of emergency line from my hip and tied into the bitter end of the floating string that was once our lifeline, then closed my eyes and searched by feel. Gooey silt and clay were jamming into every bit of my gear. I backtracked a bit and then shuffled forward time and time again until I broke through the restriction into a larger space, all while pulling, pushing, and unsticking Ruth from the ceiling protrusions along the way. I finally found another ball of string on the floor of the cave and tied my emergency reel into the mess, cutting off my reel from the now patched line. I carefully retied and stowed my spool and the cutter, knowing I would probably need it again. At times Ruth held my hand, and at other times she felt like an octopus with eight arms on all parts of my body, holding on to anything for comfort. Trying to hold my own fear in check, I kept pushing her forward toward the exit.

Without warning, she suddenly turned around to face me and headed the wrong way, back into the cave. When cave divers panic, they can lose their sense of direction. There are many sad stories about divers who split up when the shit hits the fan. In a state of hysterical terror, you can be convinced you know the way out, but instead swim deeper into a cave, toward inevitable death. The other person makes the heart-wrenching decision to continue out and save their own life. I was doing everything I could to prevent that from happening to us. I knew the right direction to swim, but Ruth was confused, and I had to convince her to reverse course, to seek the way to safety.

Fearful thoughts tried to drive their way into my head: I might not get home to my Robert; this cave would be my tomb; it would be humiliating to have accomplished so many successful extreme exploration dives only to perish in this little local spring. It was worse to think about returning to the surface without my buddy, a fresh-faced young woman with her entire life ahead of her. I took a deep breath, realizing that I had to deny those fears, for now. I had to stay pragmatic. I caught her as she tried to swim by me, grabbed her soft, cold, once confident hand now marred by cuts from thrashing at the sharp stone cave walls and steel equipment.

I held it and would not let go. We rose to the ceiling to a tiny sliver of clear water, and I tried to show her my slate, where I had written, "Calm down. Don't give up. Your gear is catching the line." I could barely see her eyes and was not sure if she could read my slate, so I placed the guide line and a plastic directional arrow pointing the way to the cave exit, to our ultimate safety, in her hand.

In a split second, Ruth was swimming away from me with all the strength she could muster, and in her haste, the silt kicked up by her fins blacked out any of my remaining visibility. I was now deep inside a fragile cave, blind, and without a buddy. I was terrified that she might not find her way out before running out of gas.

I felt around me in the darkness. I found more old deteriorated and broken line. Worse, the clay silt was clogging the regulator on my right tank. My twenty-three-year-old Omega started to free-flow, the life-sustaining gas escaping rapidly through the mouthpiece. I turned off the tank. I could have switched to the other working tank right away, but I knew I needed to save every bit of gas in the event I found Ruth. She was breathing fast and might soon be out of air, and we'd have to share whatever gas I had left. I gingerly turned the valve of my right tank on and off for each breath, sipping the blast of air and then shutting it down, trying not to waste a single bubble while I intentionally lengthened the pauses between breaths. I concentrated to slow my heart and meditated to find a rhythm that would provide a delicate balance of breathing and gas conservation. Each breath was now laced with a spray of clay and mud that coated the back of my throat.

The emotional stuff I dreaded kept rising up, but I had to concentrate to swallow it back down. I didn't want to die. And I couldn't leave Ruth. I was in charge of my fate. Ruth's fins were no longer stirring the clay into the water column and obscuring my chances for survival. I tried to remain calm and still, to allow the spring's flow to flush away the silt. Again I had to patch the line for the next divers, even knowing, in the very deep recesses of my mind, that the next team that swam into this cave might be coming to retrieve my remains. If that happened, I wanted to be sure they had a line to get out.

I knew the right direction to swim to guarantee my personal safety, but I chose to backtrack into the belly of the dark, cold cave to make sure Ruth wasn't left behind. I couldn't live with myself if I deserted her, and even though she had swum away, I wasn't certain in all the murk whether she was headed for the exit. Divers who are lost in a cave often swim into clear water, thinking that is the way to safety. But clear, undisturbed water is water you have not swum in yet. It can lead a panicking diver further into the cave, further into danger. As much as I wanted this dive to be over, I knew I needed to retrace our dive through the silt we had stirred up and would turn around to exit only once I was sure that I was not leaving Ruth behind. With each breath I carefully opened the tank valve just enough to get a mouthful of gas, then I shut it down again. It was a pace that kept me focused and gave my mind something to do other than worry.

I reached the maximum penetration point and found the water clearing up. I tidied the guide line as best as I could and turned back for the exit. It was a huge relief to again point in the right direction. Now it was a matter of sweeping the cave on my way out. The silt was settling with every minute that I got closer to the entrance. I found Ruth's sampling gear on the floor, haphazardly discarded in a side passage, and my heart sank. Worried she had bolted into the tiny space, I tied on a jump reel and searched the perpendicular tunnel. Nothing. The water was too clear for anyone to have swum here. I continued the nerve-racking swim toward the entrance, glimmers of visibility returning, but still no sign of Ruth. I looked in every crack in the ceiling pocket, hoping she might have popped up into a lucky pocket of air before she ran out of gas. It had now been more than an hour since I last saw her. If I couldn't find her, I was prepared to surface to call out a recovery team, then return to the cave with any remaining gas. I was accumulating a bit of decompression time, but had already decided to skip it and risk getting the bends to call for help.

Finally, I reached our original reel near the cave exit and strained to look up. In that moment I was most afraid, worried that I had somehow missed Ruth and that she might be dead. Then, near the surface in the fissure, I saw her. She was parked in the entrance, peeking below the waterline and praying for some sign of my return. I could see that she was crying. I had never been so happy to see anyone in my life and I know she felt the same. I needed to stay at fifteen feet to allow my body to decompress. I put my hands together as though I was praying and wrote "Thank Goddess" on my slate. She wrote back, "Called 911." As I rose up the fissure crack to the beautiful light of day, the poor dead turtle popped out of his crevice and floated up ahead of me. We were both freed from the cave's cold grasp.

On surfacing, I learned that the cave-diving cavalry she had called was already on its way. Seven close friends, including Woody Jasper, were en route to the spring with full tanks and hopes for rescuing us. They had decided not to call Robert. For seventy-three minutes, I had been dead to my friends, but they had spared my husband the immediate anguish. One hour and thirteen minutes is an eternity of waiting, suffering, and reflection.

FEW PEOPLE WILL ever walk that close to the edge, that close to their own death. It made me reflect on the meaning and purpose of my work and the risks I was taking. It also made me think more deeply about the other people who were entangled in my decisions. My decisions were family decisions. They could affect my entire community if I got myself killed.

In the following hours and days, phone calls and emails came from friends, telling me all the things they would have wished they had said to me before I had died. They were sharing what they might have said in a eulogy. The aftermath also required a time of reckoning with my military veteran husband, who suffers from PTSD. Robert had experienced more death in the first four years of our relationship than he had during his entire career as a combat photographer. He didn't want to get close to any more of my cave-diving friends. Although he recognized that cave diving made me who I was, he didn't want to go to any more funerals, especially mine. Every time I hopped in the water, I had to think about Robert too.

A few days after the scare, I received a short note from Woody Jasper. Like Woody, it was straightforward: "The mental message I was sending to you as I was driving down there was to remember the compass, and just get back to the job at hand."

Up until that point in my life, that is how I had coped with stress and loss. Just get back to the work at hand. But at some point, the stress and grief accumulate to the point where you have to deal with it.

MY DEAD FRIENDS

2012

FEW MEN ARE married to women who face danger nearly every day. Most husbands assume they will see their partner at the end of the workday. No matter how much he accepts and supports my job, Robert is still left counting the hours until I come home. Regardless of the assurances I give him, when he sees one after another of my colleagues die, it makes my career harder to defend. Almost every day he has to think about what he will do if something happens to Jill. As I head off to tough places, to test new equipment or run dangerous projects, the only thing he can do is wait.

The sobering loss of Wes Skiles and the incident at Otter Springs left me feeling rudderless and increasingly worried about diving with friends, afraid they would die or get me killed. I pulled back from teaching and became ever more discerning about the students I took on. I lessened my social interactions with cave-diving friends and turned inward. I had to question whether I was on the right path. Besides my own scare, the cumulative body count was getting too high. Images of all my dead friends came rushing in. I couldn't even keep their backgrounds straight in my mind anymore. There were too many of them.

After a friend dies, there comes a time for emotions and processing. Most of the time, I bottled up the grief and put it aside until I could process it privately. But each new death brought with it the memories of all the others. They piled up until I could no longer delay the bereavement. After Otter Springs, I fell apart. Memories flooded back to me in nightmares as vivid as the day they happened.

I saw lights swimming toward me in the cave, flashing frantically, asking for help. Eight legs, but only six were moving. The swarm was struggling as they fought to lug another diver. The victim had sunk to the floor of the cave a few minutes earlier on a vacation dive but was now bouncing off the ceiling and getting snagged on the rocks as he was dragged by his friends who were trying to save his life. One diver tried to hold a regulator in the victim's mouth, purging it fruitlessly to help him breathe. A mass of bubbles, limbs, silt, and ghastly gaping eyes. His face was familiar, but bulging, blue, purple, and red. Technicolor fluids sloshed inside his mask. Chunks of flesh spilled out of his mouth, the result of the reducing pressure as we dragged him up the chimney of the cave. Bloated and overstuffed, lips pursed outward, he was a vacant shell. If I couldn't get this group in control, we would lose more than this man. Pumping his chest in the shallow water of Little River Spring, I work to try to save his life and then I wake up from the nightmarish recollection of another unsuccessful resuscitation attempt.

I feel bad that I can't always recall their names or even the dates anymore. My dead friends are a mixture of difficult memories. Some are resolved and some are yet to mourn properly. I have seen the debossed imprint of their bodies in the white silt of the cave floor. I have dragged their lifeless remains from the water, eyes obtruding and faces frozen in translucent blue terror. I have struggled to breathe life into their lungs while they puked into my mouth. For some I wrote eulogies, called their partners, or sobbed alone at home with nobody listening.

I continue to swim through their graves.

The news does not always come face to face or over the phone. In the days of social media, we live and die in public. Robert slowly climbs the stairs with his laptop in his hands. I can tell what is coming from the weight of his steps.

"Honey," he gently calls. "I have some bad news."

I've learned that phrase and its cadence.

"Carl's dead."

I don't need a last name because I would have been seeing him in two weeks to train on some new technology.

Carl is the third colleague this week, dead on a rebreather. It is all too much to handle. Although Robert has never met Carl, I am grateful that Robert will hold me in silence and allow me to grieve in his arms.

Sometimes I learn about their deaths in the news on Facebook.

"A man who got into difficulty while diving off the South Devon coast on Saturday afternoon has died." He was Robert's friend too, and we grieve together.

"Ten people, including two who were found Monday, have had fatal incidents inside the Eagle's Nest." The headline vilifies the cave.

"Four of the divers managed to find their way back out, but the remaining four were later found dead in the tunnel." Divers' last moments are reduced to attention-grabbing story leads.

I used to get angry at my friends for leaving this world too soon, but I now know it is pointless to suspend my own life in an attempt to bring them back. I just keep saying their names so I won't forget them. I learn from their mistakes and pass on those lessons to my own students. In that sense, it restores those divers' dignity by respectfully showing that even great divers can make bad choices. Although the memories still haunt me some nights, I now choose to focus on living fully rather than marking time, frozen in place, fearing death.

Robert had faced death in his military service, and his guidance helped carry me through the loss of our friends. He helped me navigate the way back to doing what I loved when it seemed like it was all too much to handle. He told me that the very thing about me that both fascinates and inspires him is the same thing that scares the shit out of him. By acknowledging and supporting my passions, he gave me the greatest gift of a solid relationship—respect. He has no stomach for my work, but he supports me at every turn. He is proud of me. And in the bond of our 7R genes, we cherish every second together. There isn't a day that passes where we don't hold hands and say "I love you." When we gaze into each other's eyes, it is with full hearts and gratitude.

A LITTLE BIT OF MAGIC

2013

WHEN I WAS a child, my father taught me to leave my campsite better than I had found it. In that spirit, I needed to know I was leading a life of service and integrity. If I was going to break new boundaries, I wanted to ensure that I was not just courting death in some childish fascination about being an explorer. Any risk I took needed to be one that was fair to all around me. It wouldn't make a damn bit of difference to me if I died, but what about Robert, my family, or anyone who had to retrieve my body? I wanted my career to have meaning and purpose.

With Robert's support and encouragement, I realized that my expeditions were not just self-aggrandizing; they were offering meaningful contributions to science, discovery, and education. Doing a personal inventory, I recognized that I was happiest when I felt like my diving served a cause for education, outreach, or mentoring. Even though he wasn't personally interested in diving, Robert chose to take a bigger role in working with me, hoping that, together, we would commit to finding a life balance between adventure and service. At another crossroads in my career, we decided we would focus on constructive projects that helped make the world a better place. With this mission, we launched our We Are Water Project. Leveraging attention-getting stories from my cave-diving adventures, we found ways to teach people how to preserve the planet's drinking-water assets. Robert quit his dangerous and stressful job as a prison nurse, and we put our full efforts and financial savings into making an independent film about water literacy—helping people learn how to be better stewards of the water planet.

But Robert had one more simple request. He wanted to participate in a remarkable expedition, and asked me to take a break from diving so we could have our own expedition without diving, media, or distraction. I set aside four months in the upcoming summer and canceled all my diving work.

We had just finished editing our film, *We Are Water*, and still had no plans for its distribution. But we thought, what better way to share the film with the world than to do it in a carbon-neutral way? And so in a moment of either pure genius or complete folly, we decided to tour the movie on a west-to-east journey across Canada. To make it fit the ethic we were advocating in the film, we would do our tour by bicycle, without the support of a vehicle. We would camp each night and show our film in dive shops, YMCA facilities, homes, churches, and libraries. We would chronicle our journey online and so share the message with the world. It was a 4400-mile recovery-quest that reinvigorated both of us. The ride and the movie were also a salute to Wes.

When Robert and I struck out across the Rocky Mountains, we knew we were making the ride for the good of our relationship, in memory of a friend, and to unite our voices as water advocates. The ride also reminded us about our love of cycling. We had dated by bicycle, often riding thirty miles or so and picnicking at my favorite river swimming holes. We had taken trips to ride iconic trails in South Dakota and southern Ontario. When my van was not packed with dive gear, it had two dirty bicycles in the back.

The ride turned out to be one of the most physically demanding and psychologically challenging things I had ever done, and it was just as hard for Robert. We rode about sixty miles each day, in snow, rain, wind, and hot sun. We fell. We got each other back up. We were simultaneously sore, exhausted, defeated, and exhilarated. We stood on the peak of Rogers Pass in a screaming blizzard and realized that we had to keep riding into darkness to get to a safer place. In Cochrane, Alberta, after a bad wipe-out that probably should have landed him in the hospital, Robert got right back on his bike. I suffered through pneumonia *and* sunstroke, enduring a high fever and blistering thighs that were sunburned right through my cycling shorts.

But despite the pain and challenge, it was also the most intimate experience I can imagine a couple sharing. Each night, we pitched our tent in some new place, sometimes in a community park, other times in a bear-infested forest. We rolled out our meager camp mats and snuggled to stay warm or lay naked in the relentless summer heat and mosquitoes. We tended to each other's wounds, massaged each other's tired legs, and talked until we fell asleep.

By the time we dipped our toes in the Atlantic Ocean in St. John's, Newfoundland, we had a new lease on life. For both of us, the trip was about finding the wonder and connection again. And, perhaps equally important, the shared suffering along the journey solidified our already close relationship. It strengthened our commitment to bring the best people we could be to our marriage. As Robert heard me make presentations in town halls and museums, he listened to me speak about "swimming through the veins of Mother Earth." He heard the reverence in my voice and my love for the water planet. He realized that I loved my job almost as much as I loved him, and that, as dangerous as that job was, it was what made me who I was, and who he loved. He would not come between me and my water world. He would love and support me even though my work scared him to death.

In diving, the wonder comes in the most remarkable places. For me it has arrived in the magic of a cave filled with ice formations beneath the Ural Mountains in Siberia. It has happened at the moment I plunged into the water for a sunset swim across the Arctic Circle. It has flown on the wings of a cormorant that dived beneath the surface and swam in front of my camera. It came in the joy-filled playfulness of a pack of Steller sea lions that pulled on the fabric of my dry suit and nipped at the flashy dive computer mounted on my wrist.

The wonder-filled moments of diving are everywhere if you pay attention. I have found that when I share the wonder, I can transform the thoughts and opinions of those around me. Whenever the challenge of defeat or the stress of loss has overwhelmed me, I remember the wonder that brought my colleagues and me to this endeavor. When I feel trapped in the grief of losing another friend, I remember the moments submerged among a hundred humpback whales off Newfoundland or swimming with a school of massive mobula rays in the Azores. I remember the tiny ancient bat sealed beneath a sheath of calcite rock and the prehistoric bear skeleton in the depths of a Mexican cave. I remember collecting rare new animals from inside lava tubes in the Monte Corona volcano and tramping through an Egyptian sandstorm to dive into an oasis. I remember frilly purple coral on the Challenger Plateau at 450 feet and glow-in-the-dark sea pens off British Columbia's coast. The beauty and magic make the heartbreak of grief a little easier to manage.

THE NEXT FRONTIER

2017

ON A HOT fall day in November 2017, I reconvened with Dr. Bill Stone and colleagues from his Texas-based aerospace company. In the years since Huautla and Wakulla Springs, Bill and I had kept in close contact. While I have continued to dive deep into the earth, he has advanced the technology that helps us extend our range beyond where I or any other human will go. His ultimate quest was never to equip technical divers. His interest in underwater caves was as a proving ground for work in outer space. His rebreathers are for space walks. His mapping vehicles are destined for the great planets and moons beyond.

At a dive site fittingly called Wes Skiles Peacock Springs State Park, we set up an unlikely mission control center. Engineers propped up white folding tables covered with computer monitors and high-end laptops. Pale-skinned programmers slathered on sunscreen and bug spray under the canopy of sturdy maples and oaks. A barred owl called through the air that was thick with the buzzing of cicadas, and Bill Stone stepped into the group to offer a briefing.

His face was withered and lined from years of stressful work, but his eyes provided a strength and conviction that was still infectious. "We're going to make history today," he declared. "Today will be the first time anyone has put a robot into a 3-D underwater labyrinth for the purpose of exploration. It has never been done."

I was excited to be a part of this momentous day in diving history. The giant mapping device I had driven at Wakulla Springs was the grandfather of a sleek extraterrestrial explorer called Sunfish. The difference was that Sunfish ultimately will no longer need my help or that of any other diver. The artificially-intelligent robot swimmer was prepared to enter the cave and map on its own.

I swam into the cave to set up large underwater lights so that I could film the historic mission for television. The flat orange aquabot whirred and pulsed but remained poised like a runner in the starting blocks, hovering in my bright lights and holding in place. A light on the front flashed five times to let me know it was about to launch. As I hit Record on my camera, my heart skipped a beat. The device rotated and pitched as if craning its neck to look around. Using its thrusters, it swam decisively to a rock pillar while I chased with my camera. It rotated and spun again, then made a choice to launch to the left, away from the exploration line left by divers. Step by step, it built its own digital model of its surroundings and confidently proceeded to explore the underwater cave passages. For me, watching it learn and move through the tunnels was like witnessing the creation of Adam.

I didn't fear being replaced. I was exhilarated, transcending the limits of earthbound existence. Sunfish would soon be able to dive deep beyond anywhere I could ever go.

I felt a brisk chill of vitality tingling on the back of my neck as I chased Sunfish on its path through the cave. Beside me Paul Heinerth held a large cinema light. I gently touched him on the arm to direct the light where I needed it. "We dived together at Wakulla. It seems so right to share this moment today," I thought to myself. In the years since our divorce, we had nurtured a better friendship and mutual respect for each other and continued to work as colleagues in our small community. Somehow, without the pressure of marriage, it was easier to be friends and remember the best parts of what we had brought to our relationship. I had called Paul to volunteer as a lighting assistant for the shoot, trying to bring in as many of the original Wakulla team members as possible to see the ultimate fruits of our labor almost twenty years later.

We assigned Paul to a significant task. While I ran the video camera, he would disconnect the robot's tether and release it to explore unaided—the first completely robotic AI cave explorer, with a mind of its own. Paul lined up alongside Sunfish while it hovered in place, firing small thrusters in several directions. He shifted his position to get beside the cable that was trailing from the starboard rear motor shroud. He grasped the yellow strand where it met the stainless steel quick-release plug. When I nodded from behind the camera, he gave it a gentle tug and broke the connection.

For a brief moment I waited for the robot to move, and then the motors sprang into action and Sunfish swam away all on its own. I chased it with the camera while it recorded a long string of values into computer processors inside its brightly painted hull. Invisible to us, it was pinging the walls in rapid succession and measuring the distances and its precise location and orientation. Sunfish swam at a nimble pace, then rotated to capture its place within the voids of the complicated tunnel. Pausing, it appeared to think about its next sweep into the unknown. Moments later it would float forward, spin again, and continue filling in the blanks of its digital map.

It seems strange to get choked up over a robot, but for me, it was a recognition of the impact of my work over the decades. I was just a small cog in a wheel, but this was full-circle stuff. I knew I was contributing to exploration in my brief time on earth. I felt like Sunfish, loaded with twenty years of wisdom from everyone along my journey, now suddenly completely free of all burdens, disconnected and exploring my own direction.

The technology in this aquanaut-robot would not just describe the places I would never see; it had a much bigger mission. Stone's life work was focused on discovering life beyond our world. On Jupiter's moon Europa, aided by a nuclear melt probe, Stone's robot might eventually pierce through the ice-covered surface to map the liquid ocean beneath. I might never make it to space myself, but a cave diver, if only a robot version, will be the first extraterrestrial undersea explorer. And I will know that I played a part in its development.

I have also evolved to realize a new maturity about my past relationships. Paul and I had once served a purpose in each other's lives. There was no sense replaying the difficult parts of our marriage. I learned to accept the good, bad, and ugly moments in life, since they had all contributed to who I am today. By releasing resentments, accepting our life as it was, and forgiving each other, we could be friends and colleagues.

This day felt momentous. It was only an instant in the service of discovery, but felt enormous in my personal growth. Around the campfire more than two decades ago, in the depths of the Sierra Mazateca mountains, we dared enough to dream and let our imaginations soar, even with the possibility of defeat looming nearby. Neither Stone nor I could have ever predicted that our lives would play out as they had. But despite the challenges, the failures, and the losses, we just kept making the next best step we could to move forward. I could have quit after getting bent or given up when harassed by a bully. Bill could have walked away from caving when Ian Rolland died or bailed on his ideas that people said would never work.

But when we transcend the fear of failure and terror of the unknown, we are all capable of great things, personally and as a society. We might not always know where the journey will lead us. We might feel a burden of difficulty, but all paths lead to discovery. Both good and bad life events contribute to the fabric of who we are as individuals and as a civilization. If we continue to trek purposefully toward our dreams, into the planet and beyond, we just might achieve the impossible. In a similar way, both Bill and I knew that our work would leave the world a better place.

EPILOGUE

White Island, Nunavut, 2018

THERE IS A torn-up, foul-smelling carcass of something white on the floor of the hundred-square-foot shanty that will be home to my five film crew colleagues and me for the next two weeks while we search for walrus, narwhal, and polar bears. A tangle of leathery hide, soft fur, and ivory bones was once a noble Arctic fox. But according to my guide, he has been ripped apart by either a wolf or a wolverine that, like me, saw this shack as a safe place. The rustic wind-bleached shed on the beachhead looked inviting after the long, wintry ride in the twenty-four-foot Moosehead canoe. I pause for a moment in a thick shroud of mosquitoes to wonder if I should try to save the fur, because somehow we have gotten to this remote Arctic outpost without our sleeping bags. They're back in Naujaat, somehow forgotten in the supply shed. As a result, tonight I'll be snuggled in my parka, lined up with my five male colleagues on a wooden plat-form like nested spoons in a drawer. Our guides are crammed together in a small tent inside a perimeter of electrified will hopefully deter the roving polar bears that are visit our camp. A brave but diminutive terrier is tie leash outside the wire fence, ready to bark as the intruders are coming. We must remember that w food chain in this stark and uninhabited place.

My diving partner and fellow documentarian Mario Cyr has been filming in the Arctic wilderness his entire career. He was the first person to photograph wild polar bears while diving, and is briefing me on safety issues. "It is very, very dangerous," he says in his thick French-Canadian accent. "You must not be close to the ice or shore or he will grab you and pull you up. Then it is all over. Just dive deep, as fast as you can if he comes at you." He finishes with a sinister laugh. "I will show you." But as I take in the warning, I can't help but note that the wide-angle lens on my camera will offer the best image when the bear is about five feet away. The thought makes me tremble.

"What about the walruses?" I ask. And he begins another quick lesson.

"No! The bulls are even more dangerous! If you get cornered in a bay or against the rocks or ice they will kill you."

As excited as I am about filming these rare marine mammals, I have to wonder whether I have bitten off more than I can chew, so to speak. I don't relish finishing my life as some animal's chew toy.

But I have learned over the years that as terrifying as the situation might appear to be, I can trust myself to move cautiously closer to the experience, one small step at a time. I'll carefully prep my fifty-pound underwater camera, get dressed in my insulating dry suit, and weigh myself down with forty-six pounds of lead. When Mario yells, "Okay, go!" I will make the final decision to jump into the bracing water or stay on the boat. The choice will be mine alone.

Sure, I could relax into a comfortable cocoon of retirement, but I would miss the opportunity to engage in the spine-tingling experiences that make me who I am. So, with the hair standing up on the back of my neck and cold sweat dripping into my diving mask, I leap into the cold water. In this wild and almost unimaginable situation, I continue to blossom in the purity of unhindered oration. I'll be afraid, but I'll never concede.

ACKNOWLEDGMENTS

This book could not have happened without the support of the love of my life, Robert McClellan. It takes a special person to be married to a woman who frequently runs off to partake in one of the most dangerous activities ever conceived. His patience, love, and guidance lift me to greater heights every day.

I am truly grateful to my parents, Bob and Mary Rabjohn, for teaching integrity, fairness, and excellence. Growing up in a supportive and stable environment with my siblings, Jan and Gord, allowed me to pursue opportunities in life along a path less taken.

I want to thank Paul Heinerth for guiding my early efforts in exploration. I owe the late Wes Skiles a debt of gratitude for helping me find the balance between my creativity and my obsession with the underwater world. And to Jim Bowden, Bill Stone, Ryan Crawford, and Kevin Gurr, the sincerest thanks for constant encouragement to follow my passion without limitations.

To fellow explorers Kenny Broad, Brian Kakuk, Barbara am Ende, Tom Morris, Annette Long, Mark Long, Phil Short, Steve Lewis, Stu Seldon, and Jeff Shirk; NYC Sea Gypsies supporters Joe Sferrazza and Renata Rojas; my big beautiful Niagara Divers' Association family; and the Newfoundland crew of Cas Dobbin, John Olivero, and Rick and Debbie Stanley—our expedition dives, exotic dinners, and exciting conversations will always be treasured.

To mentors Ernie Brooks, Sylvia Earle, Drew Richardson,

Dan Orr, Bob Evans, Margaret Tolbert, Phil Nuytten, Russell Clark, and Trisha Stovel, you are pioneers and have graciously opened doors for the next generation.

Many friends pushed me to write my story, offering early feedback and encouragement. Kristine Rae Olmsted and Pam Wooten patiently read early drafts, motivating me to see this effort to fruition. Shannon and Kenny Caraccia, Joe Heinerth, Beth and Jerry Murphy, Pete Butt, Georgia Shemitz, Scott Braunsroth, and the Skiles crew created an extended family when I needed it most.

I owe a great debt to the many friends and cave-diving colleagues who lost their lives but illuminated a safer path for all of us who follow. They are sadly missed and always remembered for their contributions and dedication to the sport.

I want to recognize John Geiger and the Royal Canadian Geographical Society (RCGS) for exploration and educational opportunities that have supported my life's goals. My association with the RCGS is beyond a little girl's dream come true, and I am honored to participate in the collaborative efforts we undertake.

Very special thanks go to Marina Jimenez, who wrote a beautiful story in the *Toronto Star* that led to my meeting Rick Broadhead, my literary agent. Rick's confident guidance has evolved far beyond a professional relationship. There is no harder-working man in the publishing business.

I'd like to recognize my editor, Bhavna Chauhan, assisted by Melanie Tutino; copyeditor Shaun Oakey; and an incredibly talented team at Doubleday Canada, Penguin Random House Canada, as well as Denise Oswald of Ecco Books at HarperCollins. They challenged me to craft a clear narrative that makes me proud. In helping me find my voice, they oftentimes left me wondering whether I was working with an editorial team or the best group of counselors ever assembled!

Numerous equipment manufacturers have provided me with the technical gear that makes my work possible. Without their

support, I'm not sure I could have afforded to stay current in my sport. Very special thanks to Santi, Suunto, Aquatica Digital, Hollis, and Light & Motion for their generosity and partnership. Fourth Element, Divesoft, Paralenz, Kühl, Aqua Lung, Waterproof, Bigblue, Finnsub, Sherwood Scuba, PSI, VR Technology, Ursuit, Dive Rite, and many others have offered gear over the years that has been gratefully used and worn to threads.

Finally, to my students and dive buddies, I'm honored to share the underwater world with you. Stay safe and keep chasing the dream!